Radio Telescope Instrumentation for Teaching

Online at: https://doi.org/10.1088/2514-3433/ae1223

AMERICAN ASTRONOMICAL SOCIETY   IOP | ebooks™

About the program:

AAS-IOP Astronomy ebooks is the official book program of the American Astronomical Society (AAS) and aims to share in depth the most fascinating areas of astronomy, astrophysics, solar physics and planetary science. The program includes publications in the following topics:

Books in the program range in level from short introductory texts on fast-moving areas, graduate and upper-level undergraduate textbooks, research monographs, and practical handbooks.

For a complete list of published and forthcoming titles, please visit iopscience.org/books/aas.

About the American Astronomical Society

The American Astronomical Society (aas.org), established 1899, is the major organization of professional astronomers in North America. The membership (~7,000) also includes physicists, mathematicians, geologists, engineers, and others whose research interests lie within the broad spectrum of subjects now comprising the contemporary astronomical sciences. The mission of the Society is to enhance and share humanity's scientific understanding of the universe.

Editorial Advisory Board

Radio Telescope Instrumentation for Teaching

Edited by
Timothy Dolch
Department of Physics, Hillsdale College
Hillsdale, MI 49242, USA

IOP Publishing, Bristol, UK

ISBN 978-0-7503-5603-9 (ebook)
ISBN 978-0-7503-5601-5 (print)
ISBN 978-0-7503-5604-6 (myPrint)
ISBN 978-0-7503-5602-2 (mobi)

DOI 10.1088/2514-3433/ae1223

Multimedia content is available for this book from https://doi.org/10.1088/2514-3433/ae1223.

Version: 20251201

AAS–IOP Astronomy
ISSN 2514-3433 (online)
ISSN 2515-141X (print)

British Library Cataloguing-in-Publication Data: A catalogue record for this book is available from the British Library.

Published by IOP Publishing, wholly owned by The Institute of Physics, London

IOP Publishing, No.2 The Distillery, Glassfields, Avon Street, Bristol, BS2 0GR, UK

US Office: IOP Publishing, Inc., 190 North Independence Mall West, Suite 601, Philadelphia, PA 19106, USA

To my wife and children, who have endured many trips to remote, low radio frequency interference telescope sites

Contents

Preface

In this work we primarily propose classroom experiments for undergraduate education, but the content will also have relevance for K-12 education and for the amateur astronomy world. The target audience is researchers working with undergraduates, or other educational professionals working with students, who already have a basic familiarity with radio astronomy and radio telescopes. The text, then, is for educators, but not a textbook for the students involved in their research projects. For pedagogically-minded faculty who wish to embark on a radio astronomy instrumentation project in a lab course or in a summer research setting, it can be difficult to know where to start —as opposed to a more common laboratory procedure like setting up a laser experiment on an optical bench. Of course, radio telescope kits exist, but frequently change and/or become difficult to acquire. For those standalone systems, detailed descriptions will be presented in the text beyond what is currently publicly available, with procedures applicable beyond the immediately showcased telescopes. Each chapter, then, forms a skeleton of a lab procedure and exercises that an instructor could develop, but with the instructor as the intended reader, not (yet) the student. Still, many individual tutorials and exercises can be directly incorporated into coursework. All chapters are not far from being developed into part of a course—those that are more open-ended are intended to help an institution build a small, permanent telescope that can be used for many projects and updated over the years as available instrumentation improves.

The book is divided into four parts:

Part I: Standalone systems for the classroom. Here, the reader will find a set of representative classroom experiments that can be reproduced locally for coursework or summer research. Each experiment outline will include examples, use cases, and how-to for the college/university classroom, with some projects still applicable in the K-12 context.

Part II: Local stations that are part of larger network or project. Some radio astronomy projects function as an aggregate instrument depending on data from many geographically distributed stations. Such cases are a unique opportunity in which (typically smaller) institutions can become involved in the context of undergraduate coursework or research projects. The author will present experiments for use in undergraduate coursework or summer research experiments.

Part III: Large facilities/missions that include student participation in instrumentation. Several world-class radio telescopes have been built in a ground-up fashion benefiting from frequent student involvement. Additionally, some student-involved radio telescopes interface heavily with other projects, sometimes space-based missions. These chapters lay out classroom experiments that utilize larger facilities in an accessible manner.

Part IV: Innovative methods to use across various radio astronomy teaching projects. In addition to examples of specific projects, some researchers have developed methods that can be used in many different facilities. We highlight specific software and hardware tools used widely in the field and propose introductory classroom exercises useful for students.

Before embarking on this tour of (mostly) home-built radio astronomy projects, as the editor, I would like to say that I am extremely grateful to the many students with whom I have worked on radio instrumentation projects over my career: RadioJove, The Low-Frequency All-Sky Monitor, the LWA tutorials, DLITE, and DSPIRA. Without working alongside you at Oberlin College, Hillsdale College, and Lorain County Community College, we would still be in front of our laptops all day (although we still are for a majority of the day). In addition to the co-authors of this book, I am grateful to the mentors I have had in learning radio instrumentation and data processing: Henry Valentin, Harvey Kaufman, Dan Stinebring, Teviet Creighton, Brent Cole, Tom Hagen, and especially to my father, when we were building strange-looking antennas in my childhood backyard. Most of all, I am grateful to my wife and children for their constant love and companionship, and, over the years, for their toleration of our journeys to many remote, radio-quiet locations in West Virginia, Ohio, Michigan, and New Mexico.

About the Editor

Timothy E. Dolch

Timothy E. Dolch is an Associate Professor of Physics at Hillsdale College. He received his BS from the California Institute of Technology and his PhD in Physics & Astronomy from the Johns Hopkins University in 2012. Before joining the faculty of Hillsdale College, he held postdoctoral positions at Oberlin College and Cornell University, both with the North American Nanohertz Observatory for Gravitational Waves (NANOGrav) collaboration. In NANOGrav he has chaired the Education and Public Outreach Working Group. He is also a research scientist with Eureka Scientific and with the SETI Institute. Primarily a transient radio astronomer, his research focuses on pulsars and using them as tools to detect gravitational waves from merging supermassive black holes. He is an author on 98 refereed publications and has taught a broad representation of courses including general relativity, computational physics, electronics, and astronomy. With Hillsdale students, he constructed the Low-Frequency All-Sky Monitor, an on-campus radio telescope, and the Deployable Low-band Ionosphere and Transient Experiment (DLITE) in Ohio.

List of Contributors

Mojtaba Akhavan-Tafti
Department of Climate and Space Sciences, University of Michigan, MI, USA

Joydeep Bagchi
Department of Physics & Electronics, CHRIST (Deemed to be University), Hosur Road, Bengaluru 560029, India

Kevin Bandura
Center for Gravitational Waves and Cosmology, Lane Department of Computer Science and Electrical Engineering, West Virginia University, Morgantown, WV, USA

George A. Carson
Tufts University, 419 Boston Ave., Medford, MA 02155, USA

Sarah Chastain
Department of Physics and Astronomy, Texas Tech University, Box 41051, Lubbock, TX 79409, USA
Department of Physics and Astronomy, University of New Mexico, Albuquerque, NM 87131, USA

James M. Cordes
Department of Astronomy, Center for Astrophysics and Planetary Science, Cornell University, Ithaca, NY 14850, USA

Louis P. Dartez
LIGO Livingston Observatory, California Institute of Technology, Livingston, LA 70754, USA

Timothy E. Dolch
Department of Physics, Hillsdale College, 33 E. College St., Hillsdale, MI 49242, USA

Jayce Dowell
Department of Physics and Astronomy, University of New Mexico, Albuquerque, NM 87131, USA

Joseph F. Helmboldt
U.S. Naval Research Laboratory, Washington, DC, USA

Chuck Higgins
Department of Physics and Astronomy, Middle Tennessee State University, Murfreesboro, TN, USA

Bhal Chandra Joshi
National Centre for Radio Astrophysics (Tata Institute of Fundamental Research), Post Bag No 3, Ganeshkhind, Pune 411 007, India

Jason E. Kooi
U.S. Naval Research Laboratory, NRL, Washington, DC, USA

Glen I. Langston
National Science Foundation, Division of Astronomical Sciences, NSF, Arlington, VA 22230, USA

Ashish A. Mshake
Cambridge University, Kavli Institute for Cosmology, Cavendish Laboratory, Cambridge, UK

Dhruba Saikia
Inter-University Centre for Astronomy and Astrophysics, IUCAA, Pune 411007, India

Shirsh L. Soni
Department of Physics and Astronomy, University of Iowa, IA, USA
Department of Climate and Space Sciences, University of Michigan, MI, USA

Gregory B. Taylor
Department of Physics and Astronomy, University of New Mexico, Albuquerque, NM 87131, USA

David Typinski
AJ4CO Radio Observatory, High Springs, FL, USA

AAS | IOP Astronomy

Radio Telescope Instrumentation for Teaching

Timothy Dolch

Chapter 1

Introduction

From its beginnings, radio astronomy has utilized relatively low-cost equipment to reveal an otherwise invisible sky. Since the time of Karl Jansky and Grote Reber and their first radio telescopes in the 1930s, the world's radio telescope collection has since expanded to include the Green Bank Telescope (GBT; the largest movable object in the world by some measurements) in the USA, the Five-Hundred Meter Aperture Spherical Telescope (FAST) in China, and MeerKAT (the Karoo Array Telescope) in South Africa, which is a core of the planned the Square-Kilometer Array (SKA). However, radio astronomy has retained its place as the only non-optical electromagnetic wavelength that one can conduct both from the ground and as a beginning student or amateur. Small-scale radio astronomy can be done on a small budget and can make significant scientific contributions. Additionally, it remains connected to "big" radio astronomy thanks to scalable arrays of smaller antennas. Distributed and/or expanded versions of these arrays such as the Long-Wavelength Array (LWA; USA), the Murchison Widefield Array (MWA; Western Australia), RadioJove (distributed), and the Survey for Transient Astronomical Radio Emission 2 (STARE2; distributed), have been making world-class discoveries by simply adding more and more small antennas. Such arrays are nothing new; a field of dipoles was the concept behind the Cambridge survey that Jocelyn Bell conducted, in which the first pulsar, PSR B1919+21, was discovered. However, thanks to Moore's Law, computing hardware such as GPUs and FPGAs allow for a greater versatility of distributed arrays and the large amounts of data they can collect, making large, single dishes less of the default design for a radio telescope than they were several decades ago. (Single-dish telescopes continue to have their niches; they remain ideal for pulsar searching and for radar astronomy, among other things. At the time of this writing, these needs have not been entirely filled since the collapse of the 300 m Karl Gordon radio telescope at Arecibo Observatory in Puerto Rico).

doi:10.1088/2514-3433/ae1223ch1

Advances in computing have not only improved world-class telescope arrays. They have also made small radio telescopes cheaper and more accessible to smaller venues such as college laboratory courses (undergraduate and beginning graduate), amateur astronomy clubs, and high school science projects. This is not to say, however, that home-built or small-scale radio astronomy is anything new. In the US, for example, the Society of Amateur Radio Astronomers (SARA) has existed since 1981, with significant overlap with the amateur radio community. Many radio astronomy projects are available online in addition to those discussed in this book: e-Callisto, the Itty Bitty Telescope (IBT), the Small Radio Telescope (SRT), the Affordable Solar Radio Telescope (ASRT), and several commercially available models. Larger platforms with online courses have emerged, such as high school-aimed Pulsar Science Collaboratory associated with Green Bank Observatory (GBO) and the SETI Institute's ARISE Lab. Some radio telescopes are available for students for remote use, such as the 20 m Telescope at GBO, available through the Skynet Robotic Telescope Network. However, the wealth of available material can be daunting to an instructor or amateur group that seeks to attempt an individual experiment or exercise, perhaps as one of several such procedures as part of a lab course, or perhaps as a one-off radio astronomy-focused lab as part of a larger astronomical instrumentation course. Other radio astronomy project books have been published, such as Steven Arnold's *Getting Started in Radio Astronomy*, but the target audiences are high school and amateur groups. Many of the projects in the present work will also be relevant to those audiences, but here our audience distribution peaks at the undergraduate physics student level.

Radio astronomy remains a highly accessible field in which students at many levels can become involved and make meaningful scientific contributions. However, the usage of radio astronomy instrumentation in undergraduate education is typically ad hoc. Here, we provide detailed descriptions of radio astronomy instrumentation projects that have fared well in the classroom and/or in student research experiences. The projects will be easily replicable or directly accessible to the larger community. While radio astronomy is itself a specialized field, the experiments outlined in this book will also be appropriate for courses in electrical and computer engineering. Physics and astronomy students will greatly benefit from these pedagogical procedures whether entering their respective research fields or going on to engineering careers.

The projects proposed in this book cover a variety of setups and science goals. In the case of the LWA tutorials and the introductory solar radio burst exercises, no equipment is required, with students performing data analysis on publicly available test data and eventually proposing observations. The equipment setups range from several hundred USD to tens of thousands USD in the most expensive cases. (A rough cost estimate at the time of this writing is given at the start of each chapter, but one must of course bear in mind that economic conditions and inflation can radically change the values shown). The frequencies utilized range from 10–1420 MHz. (A project one can also undertake is the measurement of radio meteor counts from passive radio reflection below 10 MHz, but in this work we only cover radio frequencies for which radio waves are mostly unattenuated. Instructors interested in <10 MHz science should also pursue the Interactive NASA Space Physics Ionosphere Radio Experiment (INSPIRE). The science goals outlined in this text cover a wide variety of astrophysical scenarios: ambient neutral hydrogen (HI) in our Milky Way, solar radio bursts, Jovian radio bursts, synchrotron emission from the Galaxy, observations of the Galactic and extragalactic point sources Cygnus A and Cassiopeia A, and with the LWA, pulsars and meteor radio afterglows (MRAs), which are (often) daytime meteor trails. In practice, a beginning instructor in this field will experience surprise at the degree to which human-made radio frequency interference (RFI) will interfere with otherwise good datataking. Unfortunately RFI, like light pollution in optical astronomy, will worsen with time due to increased usage of more of the spectrum more of the time. This will only be exacerbated by increasing pollution of the spectrum by satellite transmissions. We therefore focus on projects that will be achievable even in high-RFI environments.

Whatever projects a reader may pursue, getting to "first light" will be a journey. All readers should feel free to contact the authors[1] in this book and should join the SARA[2] and RadioJove[3] mailing lists.

[1] The editor's contact information is available at https://www.hillsdale.edu/faculty/timothy-dolch/ as well as https://tdolch.github.io/.

[2] https://www.radio-astronomy.org/email-list

[3] https://groups.io/g/radio-jove

Part I

Standalone Systems for the Classroom

Chapter 2

Solar Radiometry, HI Spectroscopy, and Galactic Kinematics Using a Small-Dish Radio Telescope

James M. Cordes

We have developed a radio astronomy component of a laboratory course taken by junior and senior astronomy and physics students at Cornell University. The goals are to gain experience in digital signal processing, data analysis, and fundamental aspects of electromagnetic radiation while also learning about the HI structure of the Milky Way. The course makes use of a 3.8 m parabolic dish. The feed antenna at the prime focus is a three-turn helix tuned to the 1.4 GHz band to enable observations of the 21 cm atomic hydrogen line. The enclosure immediately behind the feed contains a low noise amplifier and a calibration diode channeled through a directional coupler whose output passes along a cable to a laboratory just below the roof. Steering of the antenna is under computer control using of motor driven jack screws on the azimuth-elevation mount. Power and control signals for the noise diode also originate from the lab. The telescope is used to teach students about radiometry of the Sun and spectroscopy of the 21 cm line. *Estimated Cost: several $10,000s USD.*

2.1 Introduction

In this usage of a dish antenna for pedagogical radio astronomy, we utilize a parabolic dish with focal length ratio $f/D = 0.4$ (Figure 2.1). The goal is to arrive at a deep understanding of Galactic[1] kinematics and its relation to dark matter.

Students learn how to determine the system temperature T_{sys} by positioning a slab of microwave absorber material in front of the feed. By performing a drift scan on the Sun, i.e., fixing the pointing direction of the antenna so that Earth rotation will cause the antenna beam to scan across the Sun, students are able to determine the

[1] We adhere to the astronomical convention of using a capital G in Galactic and Galactocentric when referring to the Milky Way.

doi:10.1088/2514-3433/ae1223ch2

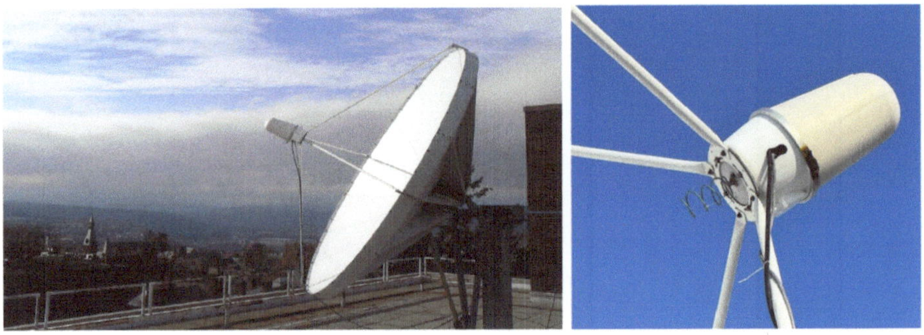

Figure 2.1. The 3.8 m f/0.4 parabolic antenna at Cornell University on top of the Space Sciences Building and the helical feed at prime focus (right panel). The azimuth-elevation mount is steered with jack screws under computer control. The front end receiver consists of an uncooled low noise amplifier and a noise diode that injects noise via a directional coupler. A cable carries the analog RF signal to a lab room while motor drive signals, noise-diode control, and power originate from the lab room.

antenna beam width (angular width of the point-spread function). They also determine the radiation temperature of the Sun at 1.4 GHz.

Spectroscopy is done by using digitally sampling the signal voltage using a software defined radio (SDR) unit to provide baseband samples. The power spectrum is calculated from these samples using the fast Fourier transform (FFT). Motivation for use of the FFT is based on the familiarly students have with Fourier series. Interpretation of spectra requires an understanding of the effects of noise and Doppler shifts caused by systematic and random motions of atomic gas in the Milky Way.

2.2 Instrumentation for Radiometry and Spectroscopy

Figure 2.2 show the essential components of the system in a simplified block diagram. Not shown is a noise calibration diode that injects a controlled signal into the signal path using a directional coupler that also accepts the output of the low-noise amplifier (LNA). Also not shown are additional amplifiers that compensate for insertion losses from other components.

The heterodyned receiver has a first mixing stage that selects a lower sideband centered on an intermediate frequency (IF) of 260 MHz.[2] With a local oscillator frequency (LO) of 1160 MHz the IF frequency corresponds to a sky radio frequency (RF) of 1420 MHz. After mixing and additional amplification, a signal splitter (not shown) provides IF signals to a commercial power sensor and to an SDR used to generate complex digital data. Both signal paths transfer data to a Linux-OS computer via Universal Serial Bus (USB) ports.

Figure 2.3 shows how the signal is modified by the first mixing stage in terms of its representation in the frequency (or spectral) domain. The narrowband signal

[2] This arrangement was chosen because a previous generation receiver made use of surplus hardware from the Arecibo Observatory in Puerto Rico.

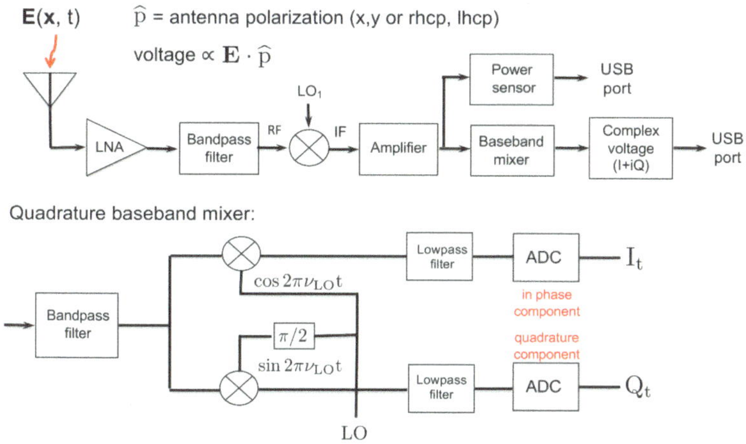

Figure 2.2. Simplified block diagram of the telescope system. A single polarization (circular polarization) is selected by the feed antenna and the antenna voltage is amplified by a low-noise amplifier (LNA) before traveling on a long cable to a receiver rack in the lab, which contains the remainder of the receiver system. Signals are mixed with a 1160 MHz local oscillator signal to shift from a center radio frequency (RF) of 1420 MHz to an intermediate frequency (IF) of 260 MHz. The IF signal is replicated with a power splitter (not shown) to provide inputs to a power sensor for radiometry and to a baseband mixing system for spectroscopy. The power sensor is a commercial device from *MiniCircuits.com* that outputs digital power data via a USB port. A software defined radio (SDR) unit is used as a programmable baseband mixer and sampler that provides complex (in phase and quadrature) data to another USB port. The circled Xs represent mixing.

selected by the receiver, LNA, and filter has a bandwidth $\Delta\nu_{\rm RF}$ and components at both negative and positive frequencies are required to represent the voltage as a real quantity. The middle part of the figure shows the spectrum after mixing (i.e., multiplication) by a real LO signal $\propto\cos 2\pi\nu_{\rm LO_1}$, which yields upper and lower sidebands. A subsequent filter removes the upper sideband (in the example shown) to yield the real IF signal centered on $\pm\nu_{\rm IF}$. Baseband mixing, as shown in Figure 2.2, yields a bandpass centered on zero frequency and it is a complex signal that requires real and imaginary parts designated in engineering applications as in-phase (I) and quadrature (Q) components.

The baseband signal $\varepsilon(t)$ contains the same information that is in the narrowband RF signal, which we write as $E_\Delta(t)$. An expression that relates the two demonstrates the relationship,

$$\varepsilon(t) = \text{constant}\times\Re\{e^{2\pi i\nu_{\rm RF}t}E_\Delta(t)\}*\text{LPF}(t) \qquad (2.1)$$

where $\Re$ designates "real part of", the asterisk denotes convolution, and LPF(t) is the time-domain representation of the low-pass filters shown in Figure 2.2. Consequently, we can analyze the power in the baseband signal or its spectrum as a means for getting the same quantities for the RF signal. If we denote the Fourier transform (FT) of the baseband signal as $\tilde{\varepsilon}(\nu)=\text{FT}\{\varepsilon(t)\}$, the spectrum is $S(\nu) \propto |\tilde{\varepsilon}(\nu)|^2$ and the power is the integral over the spectrum. In practice Fourier transforms are calculated from digitized data using the FFT.

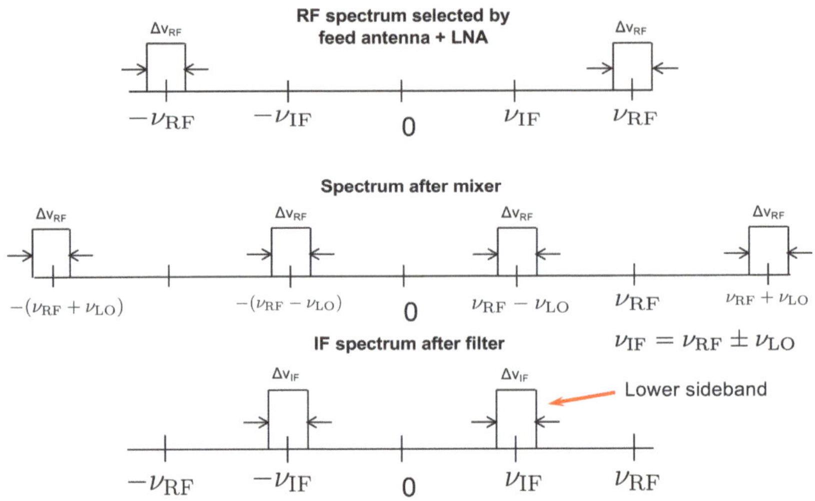

Figure 2.3. Graphic showing the manipulation of a narrowband signal at different stages in a heterodyned system. Top: narrowband signal selected by a receiver, showing signal components at negative and positive frequencies centered on $\pm\nu_{RF}$. These are present because the selected antenna voltage is real. Middle: spectral bands after mixing with $\cos 2\pi\nu_{LO}$, which creates upper and lower sidebands at $\nu_{IF} = \pm(\nu_{RF} \pm \nu_{LO})$. (Note that the LO signal referred to here is labeled LO_1 in Figure 2.2.) Bottom: IF signal after removing the upper sideband with a bandpass filter. A baseband mixing system shifts the IF signals to zero center frequency. Representation of the signal requires in-phase and quadrature components, which can be considered the real and imaginary parts of a complex signal.

2.3 Solar Drift Scan: Beam Size and System Temperature

A drift scan of the Sun along with ancillary measurements can be used to determine the system temperature of the telescope, the beam width of the antenna power pattern (a.k.a., point-spread function), and the radiation temperature of the Sun. At radio wavelengths, the radiation temperature of the Sun is larger than the photospheric temperature $T_{\odot,\mathrm{pho}} \approx 5800$ K because non-thermal radiation adds to that of the effective blackbody temperature. Departures from blackbody temperatures depend on strongly on frequency (Zirin et al. 1991), being much larger at frequencies below 1 GHz than above, and also strongly dependent on the solar cycle, e.g., as monitored at 2.8 GHz by NOAA.[3]

Figure 2.4 shows a 90 minute drift scan obtained with the 3.8 m antenna in 2019 November 19 (modified Julian data 58806.6) at 1.4 GHz. Power in the IF signal was measured at 1 s intervals with a power sensor. Noise was injected for 3 min near the beginning followed by handheld placement of microwave absorber material in front of the feed antenna. The figures shows relative power on a scale that includes zero power level. Assuming the radiation temperature of the absorber equaled the ambient temperature (~280 K), the system temperature and the nominal temperature response to the Sun can be calculated. The effective temperature from the diode can also be

[3] https://www.swpc.noaa.gov/phenomena/f107-cm-radio-emissions

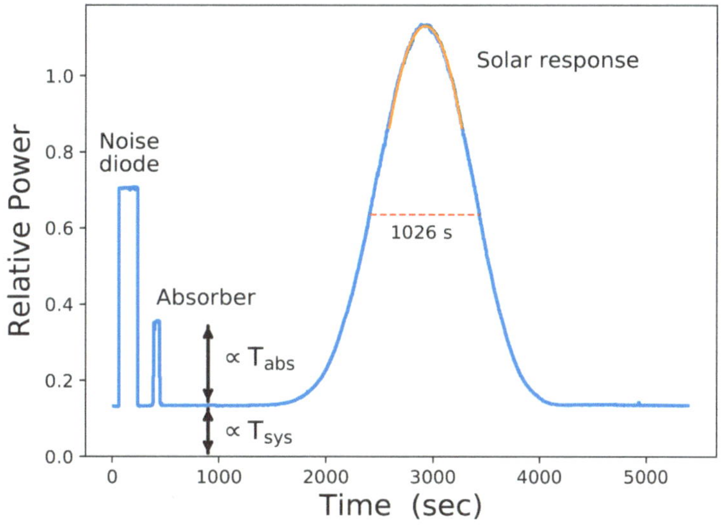

Figure 2.4. Drift scan of the Sun at 1.42 GHz using a 3.8 m antenna. The power was measured using a Minicircuits PWR-4RMS Power Sensor and the time series was normalized to unit maximum at the peak solar response. The injected noise diode signal was on for 3 min followed by the response to microwave absorbing material held in front of the helical feed antenna. The increase from the absorber is proportional to its physical temperature and the baseline level is proportional to the system temperature. The width of the solar response at half maximum is indicated. A parabolic fit to data near the peak of the solar response is also shown.

estimated for later calibration procedures (with no further discussion here). For the scan shown, the results are $T_{sys} \approx 175$ K and $T_{\odot,nom} \approx 1285$ K.

The width of the solar response in Figure 2.4 reflects the time taken for the antenna power pattern to sweep across the radio-emitting disk of the Sun. The photospheric size of the Sun $\theta_{\odot,photo} \sim 0.5$ deg while the radio size can be much larger from coronal emission. At 1.4 GHz, however, the size is still small enough that, in comparison with the beam size, the Sun can be considered a point source to good approximation. Using the temporal width of the response measured at full width at half maximum (FWHM), $\Delta t \approx 1026$ s, and the Earth's rotation rate $\Omega_{\oplus,spin} = 15° \text{ hr}^{-1}$, and the Sun's declination $\delta_\odot \approx -19.2°$ on the date of observation, the beam width is

$$\theta_{FWHM} \approx \Omega_{\oplus,spin} \Delta t \cos \delta_\odot \approx 4.04 \text{ deg}. \tag{2.2}$$

This is close to the beam size calculated for an Airy disk (the basis for the Rayleigh criterion for angular resolution), $\theta_{FWHM,Rayleigh} = 1.22\lambda/D \approx 0.2 \text{ m}/3.8 \text{ m} = 3.86$ deg. The measured beam width is about 5% larger due to the helical feed antenna's beam shape, which under-illuminates the 3.8 m dish.[4] The slight under-

[4] The term under-illumination refers to non-uniform coverage of the primary reflector (the dish) by radiation from the feed antenna if it were used as a transmitter. Severe under-illumination results if the beam pattern of the feed antenna yields a footprint on the dish much smaller than the dish diameter. This causes the beam width to be larger than expected for the nominal dish diameter.

illumination has the benefit of reducing sidelobes of the telescope's beam pattern. Note that none are obvious in the drift scan shown in the figure.

Estimating the Radiation Temperature of the Sun: The nominal radiometer temperature $T_{\odot,\text{nom}} \approx 1285\,\text{K}$ underestimates the radiation temperature by a large factor due to *beam dilution*. If the Sun's angular size were much larger than the antenna beam size, the radiometer estimated temperature equal the radiation temperature to within measurement errors (and assuming that the radiation temperature were constant across the solar disk). However, with the Sun being much smaller than the beam size, the two temperatures are related by the ratio of the Sun and PSF solid angles,

$$\frac{T_{\odot,\text{nom}}}{T_{\odot,\text{rad}}} = \frac{\Omega_{\odot,\text{rad}}}{\Omega_{\text{beam}}} = \left(\frac{\theta_{\odot,\text{rad}}}{\theta_{\text{FWHM}}}\right)^2. \tag{2.3}$$

We do not know the radio-emitting size of the Sun at 1.4 GHz for the observation date of our sample, so we will express as a ratio with the photospheric size, $\theta_{\odot,\text{photo}} \sim 0.5$ deg,

$$\begin{aligned} T_{\odot,\text{rad}} &= \left[T_{\odot,\text{nom}}\left(\frac{\theta_{\text{FWHM}}}{\theta_{\odot,\text{photo}}}\right)^2\right] \times \left(\frac{\theta_{\odot,\text{photo}}}{\theta_{\odot,\text{rad}}}\right)^2 \\ &\simeq 1285\ \text{K}\left(\frac{4.04}{0.5}\right)^2 \times \left(\frac{\theta_{\odot,\text{photo}}}{\theta_{\odot,\text{rad}}}\right)^2 \\ &\simeq 84,\ 000\ \text{K} \times \left(\frac{\theta_{\odot,\text{photo}}}{\theta_{\odot,\text{rad}}}\right)^2 \leqslant 84,\ 000\ \text{K}. \end{aligned} \tag{2.4}$$

The inequality in the last line expresses the fact that the radio-emitting size is larger than the photospheric size.

Students can compare their own estimates with published measurements, taking into account that the radiation temperature is strongly dependent on the 11 year solar cycle.

2.4 Spectroscopy of the 21 cm Atomic Hydrogen Hyperfine Transition Line

Background: The hydrogen 21 cm line is a hyperfine transition that results from interactions between the spins of the electron and proton. The energy level splitting of 5.9×10^{-6} eV corresponds to parallel versus anti-parallel spin states. The line width from gas clouds is determined primarily by *systematic motions* of hydrogen atoms rather than by the natural line width or by random, thermal motion. At $\nu = 1.4204$ GHz, 1 km s^{-1} corresponds to a 4.73 kHz frequency shift.

For Galactic hydrogen, the dominant motion is circulation of gas around the Galactic center (GC). This motion may be described quantitatively as a Keplerian orbit around the mass enclosed by the orbit. Evidently, gas is distributed in the

Galaxy so that, over a large range of Galactocentric radii, the circular velocity is nearly constant at a typically quoted $V0 = 220$ km s^{-1}. Note however that a more recent analysis favors $V0 = 240$ km s^{-1} (Reid and Dame 2016). This is referred to as a flat rotation curve. A significant amount of gas shows noncircular motion, however, especially for material very near the center of the Galaxy (i.e., within $\sim$100 pc). There are other noncircular motions associated with the Magellanic stream, the North Polar Spur and high-velocity clouds. These are features caused by explosive events in the Galaxy and by tidal interactions between the Galaxy and the magellanic clouds. Here we will describe only the circular motions associated with Galactic rotation.

Digital Spectroscopy: Spectra are obtained by Fourier analysis of the digital baseband signal. Using notation to indicate discrete sampling, the discrete Fourier transform (calculated with the FFT) of the baseband signal ε_n is $\tilde{\varepsilon}_k$ and the spectrum for an FFT length N is

$$S_k = \frac{|\tilde{\varepsilon}_k|^2}{N}. \tag{2.5}$$

Baseband samples are obtained using an SDR to provide complex data to a Linux computer and custom Python code is used to calculate the FFTs and spectra. For atomic hydrogen spectroscopy, a total bandwidth of 5 or 10 MHz is more than adequate to cover the Doppler shifted line and $N \sim 2048$ or 4096 is adequate to resolve features in the lines and, equally important, to allow excision of narrowband radio frequency interference (RFI). For a $B = 10$ MHz bandwidth, baseband samples are obtained at $\delta = B^{-1} = 0.1$ μs intervals.[5] The time span for a single FFT is then $\Delta t = N\delta t = 0.41$ ms. This is too short to yield usable signal to noise ratios (S/N), so individual spectra are averaged for 1 to 10 s (corresponding to $M \approx$ several thousand individual spectra), increasing S/N by factors of 50 to 100.

Using the 3.8 m antenna, integration times as short as 1 s are sufficient to detect the 21 cm line in directions through the Galactic plane and 10 s for directions out of the plane. An underlying explanation for this level of sensitivity is that, unlike the Sun as analyzed in the previous section, atomic hydrogen in the Milky Way is extended on scales larger than the telescope beam. This implies that *radiation* temperatures of tens of Kelvin yield similar *measured* temperatures and thus are non-negligible fractions of the system temperature because there is essentially no beam dilution.

The underlying reason can be understood in terms of the power measured from a brightness distribution $B_\nu(\theta)$ over a bandwidth $\Delta\nu$ using an antenna with effective collecting area A_e and having a normalized antenna power response $P_\mathrm{n}(\theta)$ (with unit maximum),

$$P_\nu = \frac{A_\mathrm{e}}{2} \int\int_{\Delta\nu} d\theta \, d\nu \; B_\nu(\theta) P_\mathrm{n}(\theta). \tag{2.6}$$

[5] Note that the usual Nyquist criterion expressed in the sampling theorem, $\delta t \leqslant 1/2B$, applies to real signals. The complex data we analyze require only $1/B$ intervals because two samples, the real and imaginary parts, are obtained at each time step.

For a point source of size $\ll$ the width of the power pattern, the power is approximately $P_\nu \simeq (1/2)\Delta\nu A_e \Omega_s B_\nu(\max)$, where $B_\nu(\max)$ is the maximum brightness of the source. The measured power clearly depends on the source size, $\Omega_s \propto \theta_s^2$. However, for an extended source extending over a solid angle much greater than that of the antenna power pattern, Ω_A,

$$P_\nu \simeq (1/2)\Delta\nu A_e \Omega_A B_\nu. \tag{2.7}$$

For emission at an effective blackbody temperature T_s, the power $P_\nu \simeq k_B T_s \Delta\nu$ (where k_B is the Boltzmann constant) is independent of the telescope area and beam size.[6] Though Galactic HI only approximates these conditions, its detectability is largely independent of telescope size if the power pattern does not exceed the angular thickness of the Milky Way's disk.

The S/N is determined by the peak HI line amplitude compared to radiometer noise, which we express as an rms value in temperature units, $\sigma_T = T_{sys}/\sqrt{\Delta\nu T_{int}}$, where $T_{int} = M\Delta t$ is the net integration time for the average of M individual spectra and $\Delta\nu = B/N$ is the frequency channel width for an N-point FFT. Time product $\Delta\nu \times T_{int} = (B/N) \times (M \times N/B) = M$ is independent of both the actual bandwidth and the FFT length, yielding $\sigma_T = T_{sys}/\sqrt{M}$. For $T_{sys} = 175$ K and a 10 s integration with a 10 MHz bandwidth, this yields $\sigma_T \simeq 1.1$ K, small enough to yield a high S/N for a 50 K HI line.

Drift Scans and Doppler Shifts: From this point on, we exclusively discuss HI spectroscopy that utilizes drift scans. These are conducted by pointing southward on the meridian at an elevation angle EL that corresponds to a declination $\delta = \mathrm{EL}+\mathrm{LAT} - 90°$. A full day drift scan samples the sky af fixed declination samples wide ranges of Galactic longitude and latitude.

Consider 21 cm hydrogen emission from a location in the Milky Way specified by Galactic coordinates l, b, and distance d. The measured Doppler velocity in a direction specified by a unit vector $\hat{\mathbf{n}}$ is the dot product $v_r(l, b, d) = \hat{\mathbf{n}} \cdot \mathbf{v}(l, b, d)$, where $\mathbf{v}$ is the velocity of the parcel of gas at the specified location relative to that of the telescope. As such, $\mathbf{v}$ is contributed to by Galactic rotation, peculiar motion of the HI gas relative to Galactic rotation, the solar system's motion relative to Galactic rotation, and the observatory's motion relative to the solar system's barycenter (center of mass).

A spectrum measures contributions from all positions with HI along the LoS and a telescope with normalized antenna power pattern $P_n(\theta)$ requires integration over direction as well. For a small antenna, $P_n(\theta)$ can span a considerable range of angles. However, the ~4° beam size (FWHM) of a 3.8 m antenna is well matched to the thickness of the Milky Way's disk distribution of HI, though it will average out small scale structure.

Let $\mathbf{v}_c(\mathbf{x})$ be the velocity of circular motion at a position $\mathbf{x}$ relative to the Galactic center. All objects, including gas clouds and the solar system have peculiar motions

[6] Optical systems, including radio telescopes, yield $A_e\Omega_A = \lambda^2$ (with λ = wavelength) and the brightness distribution for a black body is $B \simeq 2k_B T_s/\lambda^2$.

relative to circular motion. We mostly ignore such motions for gas clouds and make use only of the solar system's motion relative to the *local standard of rest* (LSR). We write the radial velocity at a fixed time in terms of l, b, d

$$v_r(l, b, d) = \hat{\mathbf{n}} \cdot [\mathbf{v}_c(l, b, d) - \mathbf{v}_{c\odot} - \mathbf{v}_{\odot,\mathrm{LSR}} - \mathbf{v}_{\oplus} - \mathbf{v}_{\mathrm{obs,geo}}], \tag{2.8}$$

where the Sun's circular velocity is $\mathbf{v}_{c\odot}$, its velocity relative to the LSR is $\mathbf{v}_{\odot,\mathrm{LSR}}$, the Earth's orbital velocity is $\mathbf{v}_{\oplus}$ and the observatory's velocity relative to the geocenter is $\mathbf{v}_{\mathrm{obs,geo}}$.

For drift scans on the meridian, the contribution from Earth's spin vanishes. Adopting $\Omega_{\oplus,\mathrm{spin}}$ for the Earth's angular frequency and $\Omega_{\oplus,\mathrm{orb}}$ for its orbital frequency, the radial velocity varies with time as

$$\begin{aligned} vr(\delta, d, t) &\simeq \hat{\mathbf{n}}(\delta, \Omega_{\oplus,\mathrm{spin}}t) \cdot [\mathbf{v}_c(\delta, d, t) - \mathbf{v}_{c\odot} - \mathbf{v}_{\odot,\mathrm{LSR}}] \\ &+ \hat{\mathbf{n}}(\delta, \Omega_{\oplus,\mathrm{spin}}t) \cdot \mathbf{v}_{c\odot}(\Omega_{\oplus,\mathrm{orb}}t). \end{aligned} \tag{2.9}$$

By inspection of Equation (2.9), the first dot product yields a strictly periodic variation with a period of one sidereal day because the direction specified by $\hat{\mathbf{n}}(\delta, \Omega_{\oplus,\mathrm{spin}}t)$ varies this way and the quantities in square brackets vary periodically or are constant in time. The second dot product changes with both a daily period and a yearly period, so differences in the *net* Doppler velocity are expected in different seasons. A list of contributions to Doppler velocities is given in Table 2.1 and further details can be found in Appendix 2.A.

Table 2.1. Reference Frame Motions, after Thompson et al. (2017)

Name	Type of Motion	Motion (km s^{-1})	Direction	
			ℓ(°)	b(°)
Topocentric	Rotation of Earth	0.5	—	—
Geocentric	Rotation of Earth around Earth/Moon barycenter	0.013	—	—
Heliocentric	Orbit of Earth around Sun	30	—	—
Barycentric	Rotation of Sun around Solar system barycenter	0.012	—	—
Local standard of rest (IAU Convention 1985)	solar motion with respect to local stars	19.5	57	23
Galactocentric	LSR around center of Galaxy	220	90	0
Local Galactic Standard of rest	Galactic center motion with respect to local group galaxies	310	146	−23
Cosmic Microwave Background (from dipole anisotropy)	Local group galaxies with respect to CMB	630	276	30

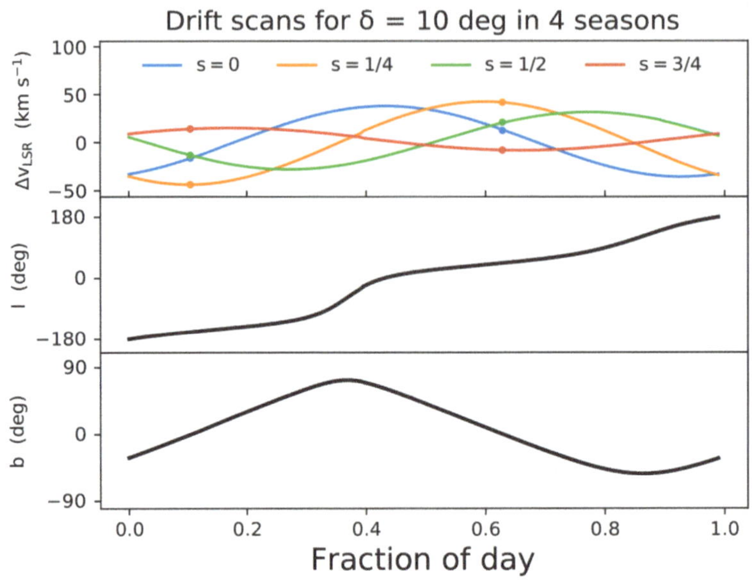

Figure 2.5. Velocities and coordinates for a drift scan pointed on the meridian at declination $\delta = 10$ deg. Top panel: Variation of Doppler shift over a day caused by observatory motion relative to the LSR. The four curves correspond to four different times of year. The two points on each curve designate values at Galactic plane crossings. Middle panel: Galactic longitude. Bottom panel: Galactic latitude.

Figure 2.5 (top panel) shows the daily variation in the total correction to the LSR $\Delta V_{\mathrm{LSR}} = \hat{\mathbf{n}}(t) \cdot [\mathbf{v}_{c\odot}(\Omega_{\oplus,\mathrm{orb}} t) - \mathbf{v}_{\odot,\mathrm{LSR}}]$ in a drift scan at declination $\delta = 10^\circ$ at four seasons designated by $s = 0,\ 1/4,\ 1/2,\ 3/4$. Galactic plane crossings occur twice per day (inner and outer Galaxy). The bottom two panels give the accompanying Galactic coordinates. A spectroscopic drift scan is then expected to show a sinusoidal variation in the Doppler shift of the HI line that adds to the systematic shift from Galactic rotation.

Drift Scan Spectral Analysis: The left-hand panels in Figure 2.6 show a 160 hr drift scan for $\delta \simeq 10$ deg as a pseudo-color graphic for spectra obtained with 10 s averages over a 10 MHz band with total integrated power plotted in an adjacent panel. Both are strongly affected by RFI that changes the mean power level and also includes persistent narrowband spikes. Variable HI emission can be seen as a narrow diffuse band that varies with a 24 hr period.

To isolate the HI line and remove changes in mean power level, a procedure exhibited in the right-hand frames of the figure is used. The top panel shows the full 10 MHz spectrum for the Galactic plane crossing at ~92 hr into the drift scan, corresponding to $(l, b) \simeq (43^\circ, 0^\circ)$. The cyan line shows the raw spectrum at full resolution with 4096 channels across 10 MHz while the black line shows the spectrum after median filtering with a block size of 10 channels. Median filtering significantly reduces the contributions from narrowband spikes. The red band indicates the frequency ranges on either side of the HI line that are used in a fifth order polynomial fit to determine the spectral baseline shape. The turquoise line

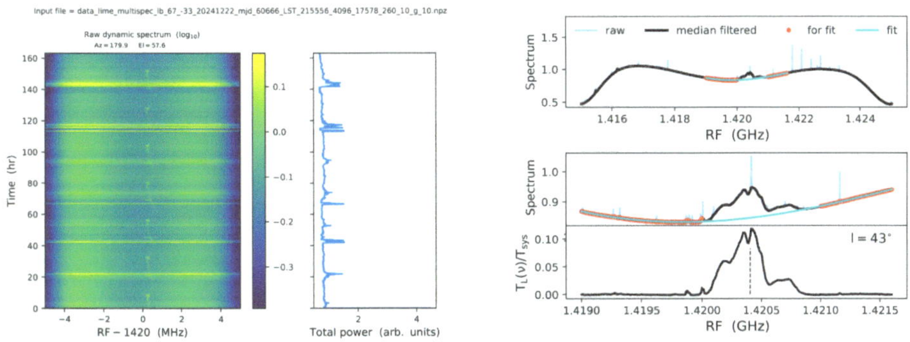

Figure 2.6. Left: Drift scan over a 160 hr span showing the spectrum in the left-hand panel and the total power (power integrated over frequency) in the right-hand panel. Diurnal variations can be seen from local activity. A large spike at the beginning of the scan is due to power injected from a noise diode. Right: Demonstration of baseline removal and spectrum normalization for a short segment selected from the drift scan at ~92 hr when the Galactic plane at $l \sim 43°$ passed through the telescope beam. Top panel: raw average spectrum for data near a Galactic longitude $l = 41°$. Also indicated are the median filtered spectrum, the portion of it fitted with a 5$^{\rm th}$ order polynomial, and the fit. Middle panel: zoom in on the spectrum. Bottom panel: difference spectrum obtained by subtracting and dividing the polynomial fit from the raw spectrum. This yields the spectral line temperature in units of the system temperature.

shows the polynomial fit. The middle panel zooms in on the central 2.5 MHz of the same curves as in the top panel. The bottom panel shows the bandpass removed spectrum calculated by subtracting the polynomial fit from the raw spectrum and also dividing the difference by the fit. Recognizing the polynomial fit as a representation of the system noise power filtered by the net system bandpass function, the result is equivalent to the HI line in temperature units divided by the system temperature,

$$\frac{T_{\rm L}(\nu)}{T_{\rm sys}} = \frac{S(\nu) - S_{\rm fit}(\nu)}{S_{\rm fit}(\nu)}. \tag{2.10}$$

This spectral flattening procedure is applied to each of the individual 60 k spectra in the drift scan, yielding the results shown in Figure 2.7 along with a Hammer–Aitoff projection of the drift scan in Galactic coordinates. The flattened spectra show much more clearly the Galactic plane crossings that occur at $l \approx 43$ and 158 deg. The inner Galaxy spectral line is broader because the larger range of Galactocentric radii encountered yield a wider range of Doppler velocities that are both positive and negative from HI gas that is receding and approaching, respectively. By contrast, the outer Galaxy spectra are narrower and mostly correspond to negative velocities.

The multi-day spectra shown in Figure 2.7 can be averaged synchronously with the 24 h period to produce an improved set of spectra. These are shown in Figure 2.8 for four different epochs, two of which are separated by about 6 months. The over-

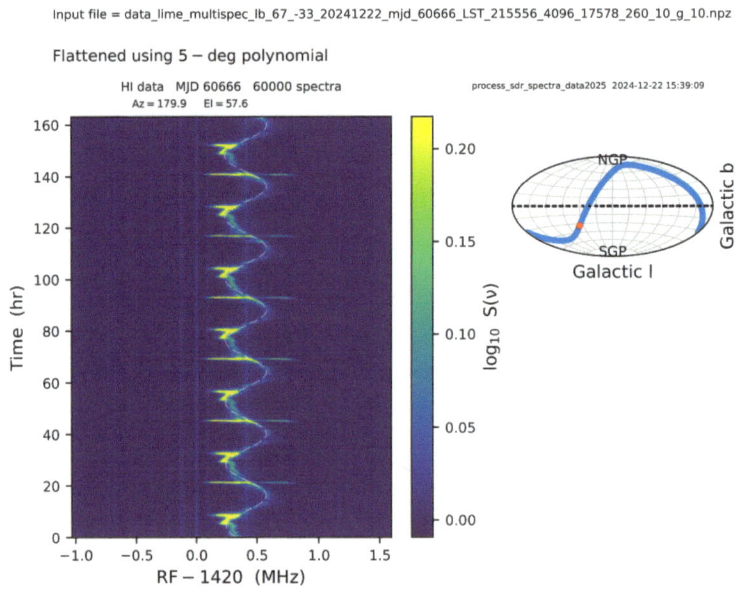

Figure 2.7. Drift scan spectra from Figure 2.6 after spectral flattening. The Hammer–Aitoff projection shows the drift scan track in Galactic coordinates.

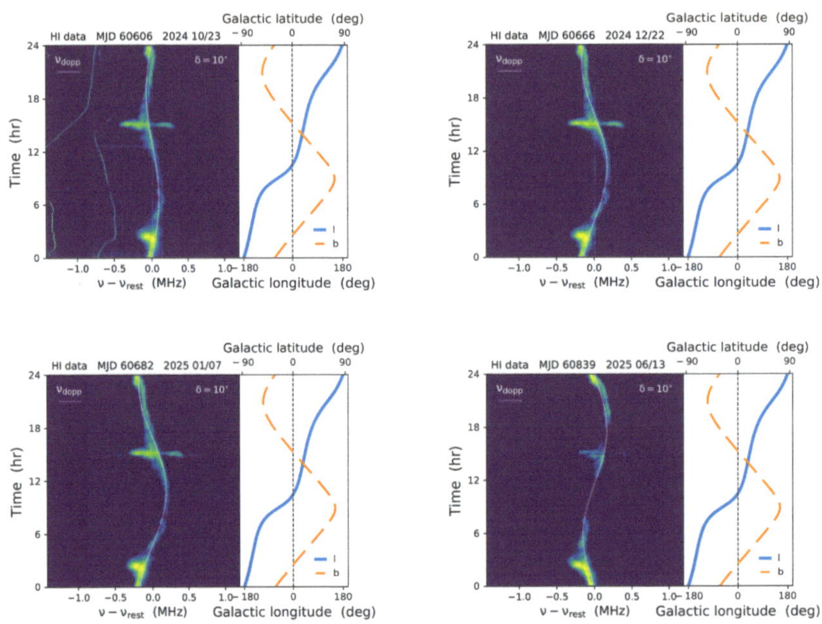

Figure 2.8. Averages of multi-day drift scans into single 24 h t or the same declination. Changes from epoch to epoch demonstrate the seasonal change in phase of the sinusoidal LSR velocity term in the Doppler velocity. Top left: 2024 Oct 23; top right: 2024 Dec 22; bottom left: 2025 Jan 7; bottom right 2025 June 13. For this scan, baseline removal was affected by stronger interference than for the earlier epochs. The ~6 month separation for the two right-hand panels correspond to oppositely directed orbital motion of the Earth.

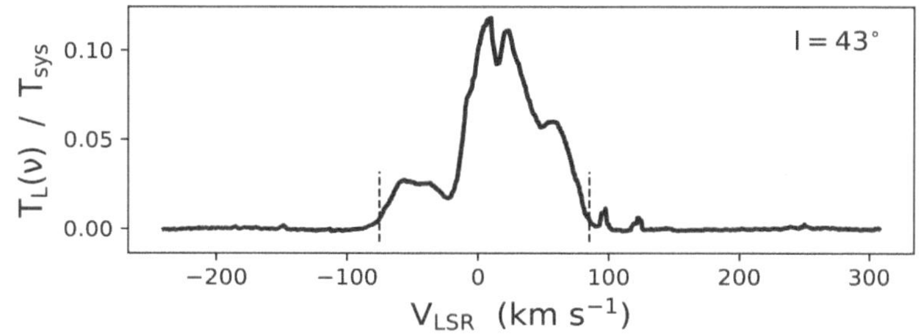

Figure 2.9. Spectrum for $l = 43$ deg versus LSR velocity. The vertical dashed lines indicate minimum and maximum radial velocities $V_{max} = -75$ km s^{-1} and $V_{max} = +85$ km s^{-1} at a level equal to 5 of the line maximum.

plotted velocity curves illustrate the daily and seasonal variation of the Doppler corrections. It is notable that the drift scan spectra illustrate Earth's spin and orbital motion as well as showing Galactic rotation.

For analyzing Galactic rotation, the spectra associated with the crossing of the Galactic plane in the inner Galaxy can be extracted. For the data from 2024 December 22, the inner Galaxy longitude is $\ell \simeq 43°$. Figure 2.9 shows the spectrum after mapping frequency to LSR velocity. Vertical dash lines indicate the extrema of the line at approximately 5% of the line peak. The associated velocities are used in an analysis later in the plane of the Galaxy's rotation rate and radial extent.

2.5 Interpretation of HI Spectra in Terms of Galactic Kinematics

In this section we interpret Doppler shifts from motions of HI gas in the Milky Way solely in terms of circular motion around the Galactic center in a flat disk of HI gas (see Figure 2.10). A flat rotation curve with constant azimuthal speed V_0 is assumed for simplicity. Of course it is well known that noncircular motions also contribute, which are due to random motions of gas clouds ($\sim$10 km s^{-1}) and systematic motions related to the Galactic bar and warping of the Galactic plane. A detailed discussion of these real-world complications and the role of dark matter in Galactic dynamics can be found in Kalberla & Kerp (2009). Circular motions are adequate for our main purpose of using HI spectra in a student lab context to estimate the rotational velocity $V0$ and the characteristic Galactocentric radius R_{max} of the atomic H disk.

Radial Velocity from Galactic Rotation: Looking down on the plane of the Galaxy from the north Galactic pole (Figure 2.11, left panel), Galactic rotation is in the clockwise direction. Galactic longitude (ℓ) is measured counterclockwise from the line connecting the Sun and the Galactic center. Thus, the GC is at $\ell = 0$ and $b = 0$ (b = Galactic latitude). For a finite size disk of atomic hydrogen, the effective path length through HI gas is a function of longitude, as shown in Figure 2.11 (right panel) for a representative disk radius.

Measurements of spectra are sensitive to Doppler shifts associated with the *difference* between the Sun's and the emitter's velocity components along the line of sight. Letting $\hat{n}$ be a unit vector pointing from the Earth to the emitting gas, and $\mathbf{v}_\odot$, $\mathbf{v}$ the vector velocities of the Sun and the gas, respectively, we define the radial velocity in this context of circular rotation (without complications of Earth and solar system motions) as

$$V_r = \hat{n} \cdot (\mathbf{v} - \mathbf{v}_\odot). \tag{2.11}$$

The sign convention is that $V_r > 0$ for gas that is moving away from us.

Now let the rotation curve be written as $V(R)$, where $R=$ Galactocentric radius (the distance from the center of the Galaxy). Since we are adopting a purely circular-rotation model, the rotation curve is thus a function of only R. The radial velocity for a subvolume of gas at radius R and Galactic longitude ℓ is then

$$V_r = R_\odot \left[\frac{V(R)}{R} - \frac{V(R_\odot)}{R_\odot} \right] \sin \ell, \tag{2.12}$$

where $R_\odot$ is the distance of the Sun from the Galactic center. We use the value $R_\odot = 8.178 \pm 0.025$ kpc (GRAVITY Collaboration et al. 2019) based on monitoring of an infrared star orbiting the $\sim$4 million solar mass black hole in the Galactic center.

Expected Features in HI Spectra: We must now consider the distribution of hydrogen gas in the Galaxy in order to estimate the expected spread of radial velocities. Hydrogen is widespread, of course, but it is localized to a thin disk that is a few hundred pc thick and several tens of kpc in radius. Within the disk, moreover, HI is clumped into small clouds, cloud complexes, and spiral arms. As a consequence, we expect to see a wide range of velocities accompanied by large variations in the strength of the HI emission.

The geometry of rotation as viewed from our vantage point yields some easily-identified features in 21 cm spectra. For a parcel of gas at distance d from the solar system and in a direction in or near the Galactic plane at latitude b and longitude ℓ, the (cylindrical) Galactocentric radius is

$$R = \sqrt{(d \cos b)^2 + R_\odot{}^2 - 2R_\odot d \cos b \cos \ell}\,. \tag{2.13}$$

The first important feature is the "tangent point," which is defined only when looking toward the inner Galaxy $-90° \leqslant \ell \leqslant 90°$. The tangent point is the position along the line of sight closest to the Galactic center and, by definition, is where the line of sight is tangent to a circle centered on the GC. Minimizing R with respect to d for $b = 0$ gives the distance to the tangent point from us,

$$d_\mathrm{t} = R_\odot \cos \ell. \tag{2.14}$$

Substituting in Equation (2.13) then gives the corresponding Galactocentric radius of the tangent point,

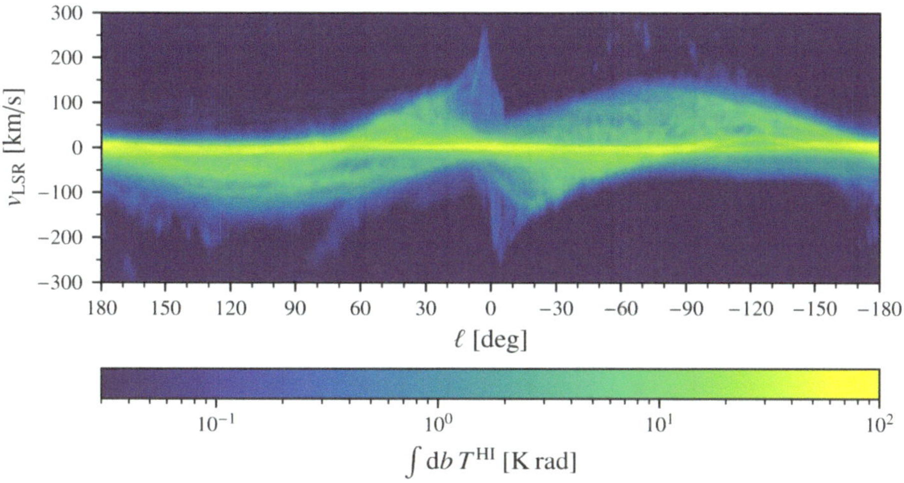

Figure 2.10. Longitude–velocity plot. Figure based on the all-sky HI survey in Söding et al. (2025). The color scale shows the HI line temperature integrated over Galactic latitude as a function of Galactic longitude and LSR velocity. Reproduced from Söding et al. (2025). CC BY 4.0.

$$R_t = R_\odot|\sin \ell|. \tag{2.15}$$

The magnitude of the radial velocity is larger at the tangent point than anywhere else along the line of sight. Combining Equations (2.12)–(2.15), the radial velocity at the tangent point is

$$V_t = V_r(R_t) = V(R_t)\mathrm{sgn}(\ell) - V(R_\odot)\sin \ell, \tag{2.16}$$

where we have used $\mathrm{sgn}(x) = +1$ for $x \geqslant 0$ and -1 for $x < 0$. For much of the Galaxy, the rotation curve is flat, so $V(R) = V_0 =$ constant, with $V_0 \approx 220$ km s^{-1} or 240 km s^{-1} in the more recent analysis referred to earlier. This implies

$$V_t \equiv V_r(R_t) = V_0[\mathrm{sgn}(\ell) - \sin \ell], \tag{2.17}$$

showing that the magnitude of the radial velocity at the tangent point is greater at smaller $|\ell|$ and that it changes sign at $\ell = 0$. From these considerations we can also determine the radial velocity as a function of distance d from us. Figure 2.12 shows the radial velocity versus d for two inner Galaxy lines of sight and one outer Galaxy direction.

The Longitude–Velocity (LV) Diagram: The LV distribution of HI is the HI spectral line versus LSR velocity plotted as a function of Galactic longitude (usually as a contour plot or heat-map). It shows the kinematics and concentration of both large and small scale structure in atomic hydrogen. Examples are shown in Kalberla & Kerp (2009) and Figure 2.10 (Söding et al. 2025) shows the LV diagram based on the all-sky HI survey (Söding et al. 2025). As an aside, we

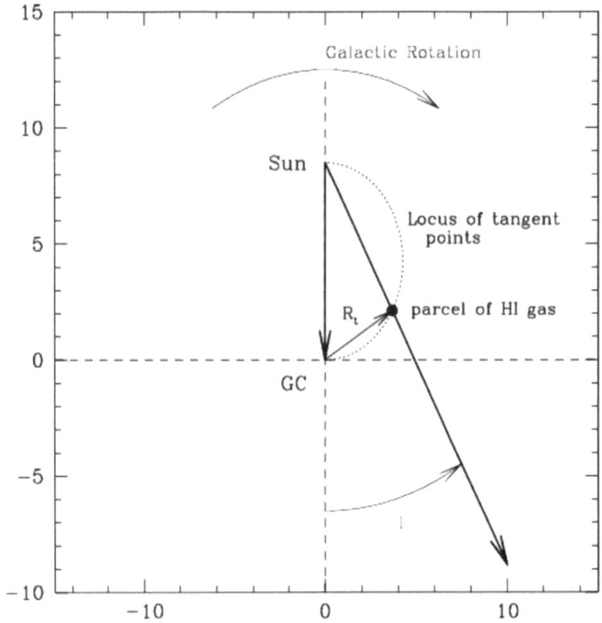

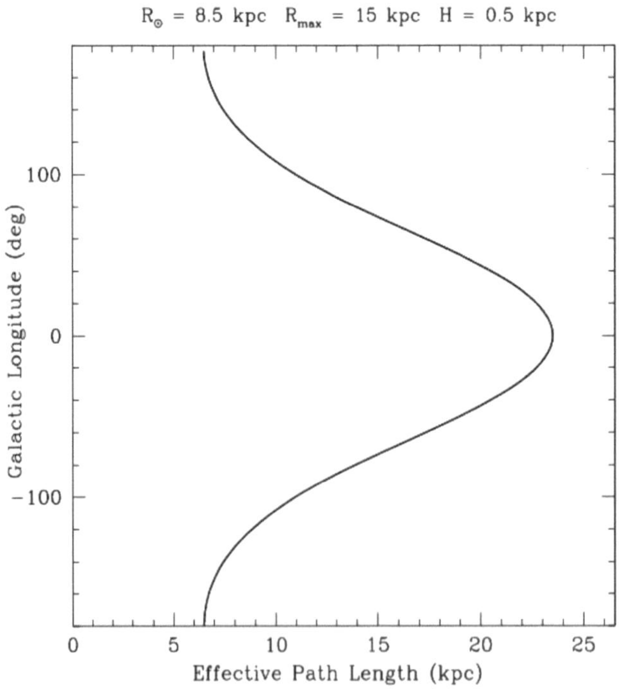

Figure 2.11. Geometry of Galactic rotation where $R_{\odot}$ is the distance of the Sun from the Galactic center and R is the distance of a subelement of rotating gas. Galactic longitude is ℓ. Right: The effective path length through the HI disk for a model galaxy with $R_{\odot} = 8.5$ kpc and a half-thickness of 0.5 kpc.

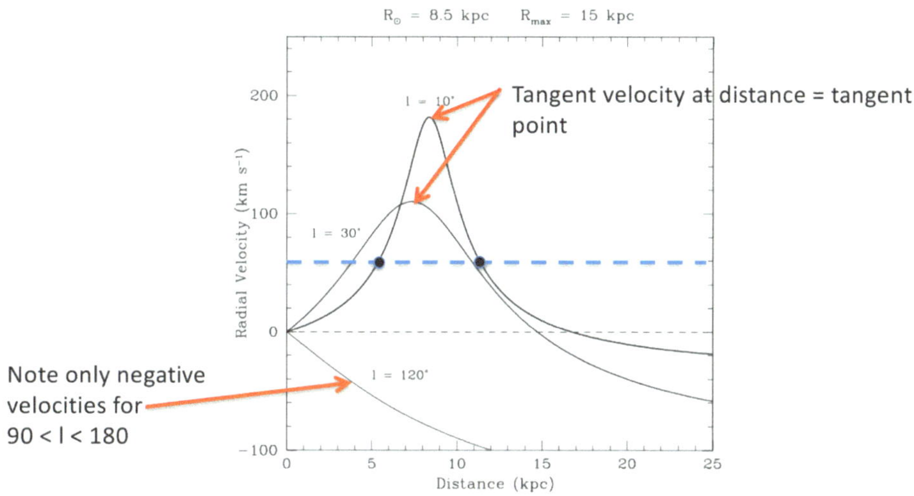

Figure 2.12. Radial velocity plotted against distance from the Sun for several lines of sight through the Galactic disk at $b = 0°$. Annotations highlight the tangent velocities for the two curves with $l = 10°$ and $l = 30°$ and the negative velocities for $l = 120°$. The dashed line and filled circles demonstrate that gas at two different distances yields the same radial velocity (the "distance ambiguity").

mention carbon monoxide, which is a tracer of molecular gas (dominated by coexisting hydrogen molecules, H_2). CO shows large-scale kinematic features similar to HI but is more concentrated to the Galactic plane and differs in its small scale structure. An LV diagram for CO may be found in Dame et al. (2001). From these examples in the literature, it can be seen that the distribution of gas (HI or CO) is very lumpy. This needs to be considered when visualizing the disk of the Galaxy and interpreting HI spectral line data.

Here we derive the elements for the HI LV diagram assuming only circular motions around the center of the Galaxy. While simplistic, it nonetheless captures salient features that are accessible to spectroscopy with a small antenna.

Using previous results we expect the following for $0° \leqslant \ell \leqslant 180°$:

1. The maximum velocity is the tangent velocity in the first quadrant and zero in the second quadrant.
2. The minimum velocity is zero at $\ell = 0°$ and $\ell = 180°$ where Galactic rotation is exactly orthogonal to the l.o.s. For other longitudes, the minimum velocity is due to gas that is farthest from the Galactic center. Thus the minimum velocity is determined by how far HI gas extends in the disk of the Galaxy.

For the third and fourth quadrants ($180° \leqslant \ell \leqslant 360°$), the two bounding curves are the reverses of those in the first two quadrants: the minimum velocity curve is the

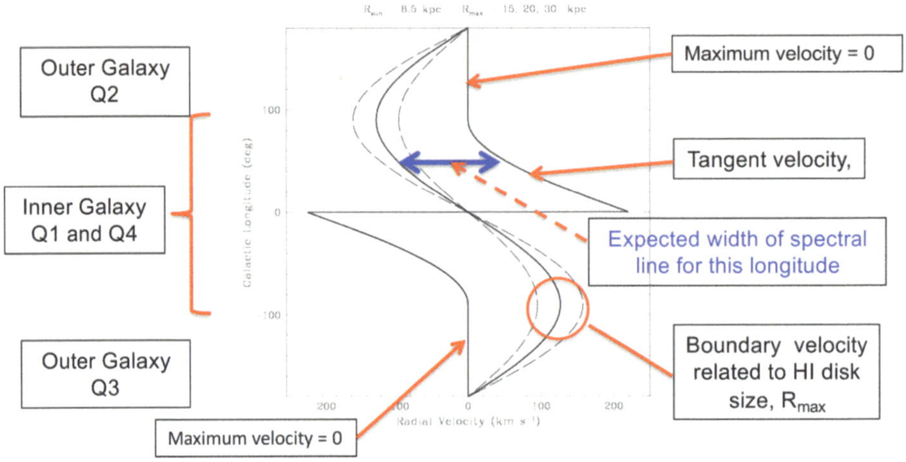

Figure 2.13. A schematic longitude–velocity plot for Galactic hydrogen that circularly rotates about the Galactic center. For the sine-wave shaped curves, the solid line uses $R_{\rm max} = 20$ kpc and the dashed lines use values of 15 and 30 kpc. Which one is which? See Equation (2.21)! Galaxy quadrants are defined in terms of Galactic longitude as Q1: $0° \leqslant l \leqslant 90°$; Q2: $90° \leqslant l \leqslant 180°$; Q3: $-180° \leqslant l \leqslant -90°$; Q4: $-90° \leqslant l \leqslant 0°$. Information about both the rotational velocity V_0 and the characteristic radius of the HI disk $R_{\rm max}$ require a spectral line measurement in either Q1 or Q4. However, only Q1 is fully accessible from the Northern hemisphere (and Q4 for the Southern hemisphere). In Q2, the maximum radial velocity (due to circular motion) is zero, so spectral lines from that quadrant cannot be used to solve for both V_0 and $R_{\rm max}$. The same is true for Q3 since the minimum velocity is zero in that case.

now-negative terminal velocity or zero; the maximum velocity corresponds to material farthest from the GC.

Figure 2.13 shows a schematic LV plot based on our model of circular motion around the GC with a flat rotation curve. Three hypothetical values for the maximum radius of the HI disk were used to make the plot (solid and dashed lines).

HI Data from the Web: HI spectral line data from the HI4PI all-sky survey of the 21 cm line survey (HI4PI Collaboration et al. 2016) can be downloaded.[7] Figure 2.14 shows an example for $\ell = 30°$, $b = 0°$ with some annotations that illustrate some of the points made above.

Determining $V0$ *and* $R_{\rm max}$ *from a Spectral Line Measurement:* Measurement of the HI line at a longitude intermediate between 0° and 90° and mapping frequency to LSR velocity yields a maximum LSR velocity $V_{\rm r,max}$ that corresponds to the tangent velocity $V_{\rm t}$, so an estimate for the rotational velocity is

$$\widehat{V_0} = \frac{V_{\rm r,max}}{1 - \sin \ell}. \tag{2.18}$$

[7] https://www.astro.uni-bonn.de/hisurvey/AllSky_gauss/

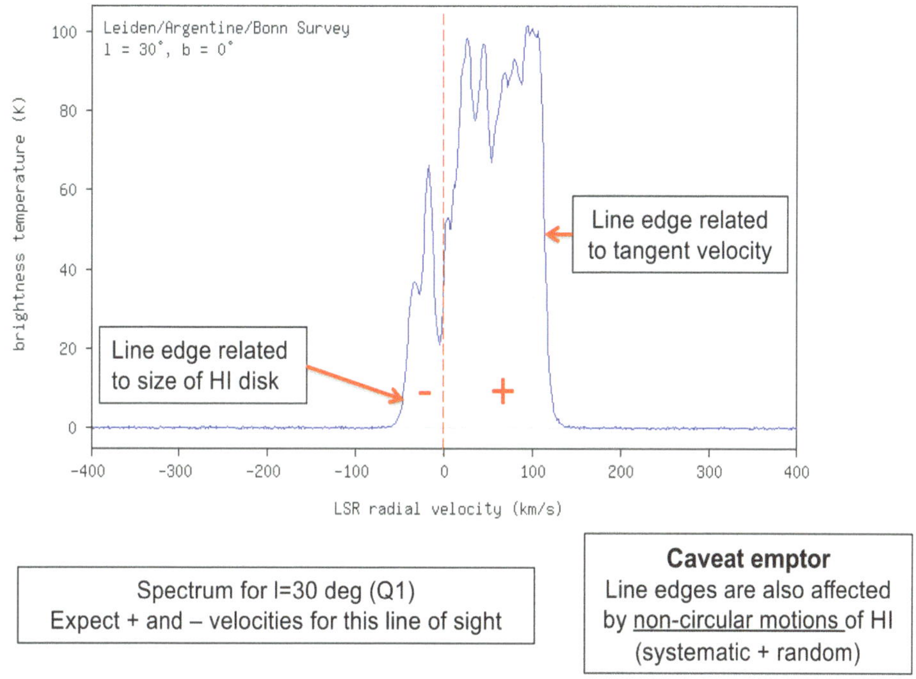

Figure 2.14. Spectrum obtained from the Leiden–Argentina–Bonn survey (Kalberla et al. 2005). Annotations indicate the information that can be inferred from the line edges.

For the disk size, suppose HI is distributed out to a radius $R_{max} > R_{\odot}$. Then, at $\ell = 90°$, the minimum velocity (for a flat rotation curve) is

$$V_{\rm r,min} = -V_0(1 - R_{\odot}/R_{\rm max}) \tag{2.19}$$

while at $\ell = -90°$ the maximum velocity is the opposite:

$$V_{\rm r,max} = +V_0(1 - R_{\odot}/R_{\rm max}). \tag{2.20}$$

For any ℓ, the bounding curve associated with $R_{\rm max}$ can be written as by

$$V_{r,\rm bound} = -V_0(1 - R_{\odot}/R_{\rm max})\sin\ell. \tag{2.21}$$

Rearranging and using the estimated $\widehat{V}0$, the radial HI extent is estimated as

$$\widehat{R}max = \frac{R_{\odot}}{1 + \frac{V_{\rm r,min}}{V_{\rm r,max}}\left(\frac{1-\sin\ell}{\sin\ell}\right)}. \tag{2.22}$$

As an example, we use the spectrum shown in Figure 2.9 for $\ell = 43°$. The caption gives minimum and maximum LSR velocities of -75 km s^{-1} and 85 km s^{-1}, respectively. Substituting these values into Equations (2.18) and (2.22) yields $\widehat{V}0 \approx 265$ km s^{-1} and $\widehat{R}max \approx 17.6$ kpc based on a Galactic center distance ~8.2 kpc. These are reasonable estimates with errors of about the 10% level. Errors are due to the rather large antenna beam size (~4 deg), interference that distorts the spectral baseline, and noncircular velocities that are not modeled in our analysis.[8]

2.A Contributions to Measured Doppler Shifts from Observatory Motions

Spectroscopic measurements are made from a non-inertial platform whose motion includes several contributions from the spin and orbit of the Earth and motion of the solar system with respect to the local standard of rest. Table 2.1 lists many of the motions that are relevant to measurements of Galactic and extragalactic sources. For measurements of spectral lines from Galactic sources, we are concerned only with motions within the Galaxy.

All corrections to Doppler velocities involve taking a dot product of the relevant velocity vector with the source direction. For Galactic HI, the three most important corrections are, in order of decreasing value:

1. **Earth orbital motion** ($\lesssim$30 km s^{-1}): this correction is periodic with a period of one year. The maximum Doppler shift of $\pm$30 km s^{-1} will be seen only for directions in the ecliptic plane.
2. **Correction to the local standard of rest (LSR)** ($\lesssim$20 km s^{-1}): this correction is constant in time (for timescales we are interested in) and for any given direction. For an arbitrary direction (l, b) the contribution of the Sun's motion is $V_{r\odot} = 19.5 \quad \text{km s}^{-1} \hat{n} \cdot \hat{n}_{V_\odot}$ where $\hat{n}_{V_\odot}$ is the direction of the Sun's peculiar velocity or

   $$V_{r\odot} = 19.5 \quad \text{km s}^{-1}[\cos b \cos b_\odot \cos(\ell - \ell_\odot) + \sin b \sin b_\odot],$$

   where $(\ell_\odot, b_\odot) = (57°, 23°)$ gives the direction of the Sun's motion using Galactic coordinates. In equatorial coordinates, $(\alpha_\odot, \delta_\odot) \approx (18.\ 0^h, 30°)$.
3. **Correction for rotation around the Earth** ($\lesssim$0.5 km s^{-1}): this correction has a period of one day.

2.B Hardware Components Used in Cornell's Instruction Radio Telescope

Here we itemize salient elements of the antenna and receiver system.

[8] In the lab course at Cornell, we ask students to do as quantitative an error analysis as possible taking these sources of error into account.

Antenna and mount: The 3.8 m f/D 0.4 parabolic reflector was purchased from Patriot Systems in 2005. The dual-axis motorized king post mount provides a sturdy platform against the elements. The mount is steerable 120 deg in azimuth and from 10 to 90 deg in elevation. Motors drive two jack screws for azimuth and elevation. Similar antennas and mounts are currently available through Challenger Communications.[9] Motors are computer controlled using a *Research Concepts* RC2000 Dual Axis Antenna Controller.

Front End Low Noise Amplifier: *Miteq* M/N AM-1477 1000–2000 MHz amplifier. The LNA is uncooled, accounting for the high system temperature. In the near future, a new ambient-temperature LNA developed at Caltech will be employed, which should reduce T_{sys} significantly.

Amplifiers:

1. Radio frequency amplifier: *Minicircuits* ZX60-3011 (400–3000 MHz) and ZFSC-2-2500-S+ (10–2500 MHz).
2. Intermediate frequency amplifiers: *Minicircuits* ZFL-500LN (0.1–500 MHz).

Mixer: *Minicircuits* ZX05-C42 (1000–4200 MHz)

Local oscillator: *Minicircuits* SSG = 6000RC Synthesized Signal Generator

Bandpass Filters:

1. Radio frequency: *K&L Microwave* 1420/T57 with 1420 MHz center frequency.
2. Intermediate frequency *K&L Microwave* 5B120-260/100 with 260 MHz center frequency and 100 MHz bandwidth.

Frequency counter: *Minicircuits* RF Frequency Counter UFC-6000

USB Power Sensor: *Minicircuits* PWR-SEN-4RMS (50–4000 MHz)

Software defined radio units: LimeSDR USB receiver[10] (up to 32 MHz bandwidth). A good alternative is the AirspyR2[11] (up to 10 MHz bandwidth). Both of these (and SDR units typically) have an analog antenna input and USB output.

The above units are used with the 3.8 m antenna system. In our lab course, students also use SDR dongles[12] with their laptops to learn the basics of signal processing and to explore the radio spectrum from about 50 MHz to 2 GHz.

Microwave Absorber: A 24 in $\times$ 24 in $\times$ 2.24 slab of AN-77 ECCOSORB absorbs microwaves with frequencies $>$1.2 GHz. This is manufactured by *Laird Performance Materials*. It is mounted on a metal panel attached to poles that allow it to be placed in front of the prime focus feed antenna when the antennas is pointed near the horizon.

USB Power Sensor: See Figure 2.15.

---

[9] www.challengercommunications.com

[10] https://limemicro.com/sdr/

[11] https://airspy.com

[12] Example units can be found at https://www.nooelec.com

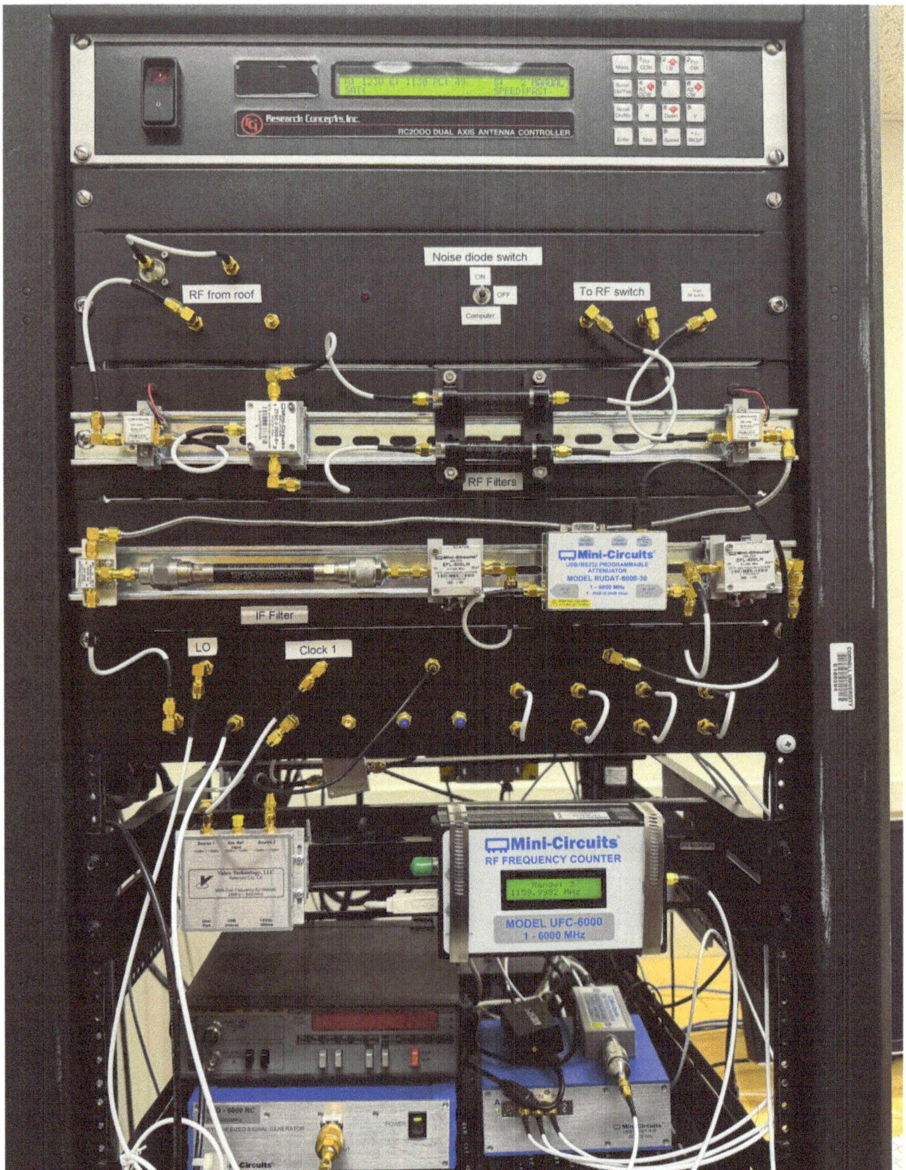

Figure 2.15. Rack for the telescope system. At the top is the motor controller. The next panel has the input RF signal from the antenna and LNA on the roof. The next two panels include amplifiers, bandpass filters, and mixer. The local oscillator signal comes from the signal generator at the bottom left and it is monitored with the frequency counter above it and to the right.

References

Dame, T. M., Hartmann, D., & Thaddeus, P. 2001, ApJ, 547, 792
GRAVITY Collaboration. Abuter, R., Amorim, A., et al. 2019, A&A, 625, L10
HI4PI Collaboration. Ben Bekhti, N., Flöer, L., et al. 2016, A&A, 594, A116
Kalberla, P. M. W., Burton, W. B., Hartmann, D., et al. 2005, A&A, 440, 775
Kalberla, P. M. W., & Kerp, J. 2009, ARA&A, 47, 27
Reid, M. J., & Dame, T. M. 2016, ApJ, 832, 159
Söding, L., Edenhofer, G., Enßlin, T. A., et al. 2025, A&A, 693, A139
Thompson, A. R., Moran, J. M., & Swenson, G. W. 2017, Interferometry and Synthesis in Radio Astronomy (3rd ed.; New York: Springer)
Zirin, H., Baumert, B. M., & Hurford, G. J. 1991, ApJ, 370, 779

Chapter 3

BHARAT Setup for HI Pedagogy: The Radiometer

Ashish A. Mhaske, Joydeep Bagchi and Bhal Chandra Joshi

We present the Bose Horn Antenna Radio Telescope (BHARAT; Mhaske et al. 2022), a low-cost and low-complexity setup that is being used in colleges and universities for the 21 cm hydrogen experiment. We opt for off-the-shelf components and Python code for easy assembly and operation. This chapter describes the instrument, observations, and calibration of the data, while Chapter 4 describes the astrophysics derived from these observations. Tutorials are provided for data collection and data analysis for all the topics discussed in both chapters. *Estimated Cost: $300–$500 USD*

3.1 Introduction

The study of the 21 cm hydrogen line, a fundamental tool for mapping the structure and dynamics of our Milky Way galaxy, was largely confined to professional observatories. Developments in electronics and computation have made the components required for these experiments low-cost. This has allowed the development of small laboratory setups for schools and colleges. The purpose of this chapter is to describe such an instrument for the detection of the 21 cm hydrogen line. In recent years, there has been a surge in interest and activity in building small, low-cost 21 cm radio telescopes suitable for amateur astronomers, students, and educational settings. Some examples of early 21 cm pedagogical telescopes are the MIT Haystack SRT (Rogers 2012; Patel et al. 2014). These projects, including the one described here, provide hands-on experience with radio technology, data acquisition, and astrophysical concepts, allowing us to detect the neutral hydrogen gas distributed throughout our galaxy and even observe the Doppler shifts caused by its motion. This has been largely fueled by advances in modern electronics and software. The advent of affordable Software Defined Radio (SDR) dongles, which were initially designed for digital TV reception, provides a highly capable and

doi:10.1088/2514-3433/ae1223ch3

versatile receiver back-end at minimal cost, replacing complex traditional radio hardware. Combined with readily available, high-performance Low Noise Amplifiers (LNAs) and simple antenna designs like horns, the core components are now easily accessible and available off-the-shelf. Open-source software platforms such as Python and GNU Radio handle the signal processing, dramatically lowering the cost and complexity of building a functional instrument capable of real astrophysical observations. The experiment was developed at the Radio Physics Lab (RPL) jointly run by the Inter University Centre for Astronomy and Astrophysics (IUCAA) and the National Centre for Radio Astrophysics (NCRA), India. (The RPL-IUCAA is now merged with the Teaching Learning Centre (TLC), IUCAA where the experiment is used for training teachers and students from schools and colleges.) Other notable projects describing instrumentation and techniques are Leech (2017), Bandura et al. (2018), Arul Pandian et al. (2022), Berkhout et al. (2023), and Fung et al. (2023). This is certainly not an exhaustive list of 21 cm projects, as their number is growing, indicating the increasing interest in this experiment for radio astronomy pedagogy.

3.2 Historical Developments

The birth of radio astronomy was serendipitous. In 1932, Karl Jansky was investigating sources of interference with radio communication. He constructed a large, rotating antenna array (a "merry-go-round") operating at 20.5 MHz. He identified a persistent, faint hiss that tracked with the stars, and correctly deduced that it originated from the center of the Milky Way (Jansky 1933). The spectral flux density unit "Jansky" (1 Jansky = 10^{-26} W m^{-2} Hz^{-1}) was formally adopted by the International Astronomical Union (IAU) in 1973 to honour Karl Jansky. Inspired by Jansky, Grote Reber, an amateur radio operator and engineer, built the first dedicated radio telescope in his backyard in Wheaton, Illinois, in 1937. Reber experimented with different frequencies, eventually succeeding at 160 MHz in 1938 and later mapping the radio Milky Way at higher frequencies (Reber 1944).

World War II spurred massive advancements in radar technology, particularly at microwave frequencies. This technology became available to scientists after the war, revolutionizing radio astronomy. Horn antennas have been used in radiometric measurements in radio astronomy due to their low side lobes, and predictable and symmetrical beam pattern. Ewen and Purcell used a rectangular horn antenna at Harvard University to detect the 21 cm hydrogen line, which is the focus of this chapter (Ewen & Purcell 1951). One can compare the difference in size of the receiver used by Ewen and the modern SDRs. Ewen installed the completed horn antenna outside a fourth-floor window of the Lyman Laboratory at Harvard, with a waveguide running through the window to connect it to the receiver and recorder inside.[1] John W. Findlay at the National Radio Astronomy Observatory (NRAO) used standard gain horns for accurate calibration of radio telescope flux densities (Findlay 1966). While attempting to characterize noise sources for satellite

[1] For images see https://www.gb.nrao.edu/fgdocs/HI21cm/ephorn.html.

communication, Penzias and Wilson detected a persistent, isotropic background noise they could not eliminate, corresponding to an antenna temperature of about 3 K (Penzias & Wilson 1965). They learned of theoretical work by Dicke, Peebles, Roll, and Wilkinson at nearby Princeton University predicting a relic radiation from the Big Bang (Dicke et al. 1965), the Cosmic Microwave Background (CMB) radiation, a cornerstone of modern cosmology. Horns are regularly used in the microwave and millimeter-wave astronomy, communication and other high frequency experiments.

Sir Jagadish Chandra Bose, working in Calcutta in the late 19th century, made remarkable and foundational contributions to the generation, detection, and study of very short electromagnetic waves, specifically in the millimeter-wave range (up to 60 GHz/5 mm wavelength). His work predated the mainstream development in radio communication (Emerson 1997). Some of his groundbreaking inventions include development of millimeter-wave spark gap oscillators and a spiral-spring detector capable of producing and detecting extremely short wavelengths. He was among the very first to use hollow metal pipes to guide these short waves – essentially early waveguides. To efficiently radiate these waves into free space and capture them at the receiver, he employed pyramidal horn antennas connected directly to his waveguide components as seen in Figure 3.1. These horns helped

Figure 3.1. Left: Sir J. C. Bose at the Royal Institution, London in 1897. In this demonstration, Bose used an early version of the modern pyramidal horn antenna and waveguide transmitter. This [J. C. Bose] has been obtained by the author(s) from the Wikimedia website, where it is stated to have been released into the public domain. It is included within this article on that basis. Right: In his later years, Bose is seen with some of his students and colleagues at the Bose Institute in Calcutta. This rare photograph features several eminent scientists: Jagadish Chandra Bose (seated at the center), Satyendranath Bose – renowned for his work on bosons and Bose–Einstein statistics (standing, second from left), astrophysicist Meghnad Saha – known for formulating the Saha equation of thermal ionization (seated, left), and physicist Debendra Mohan Bose (standing, third from left), a pioneer in cosmic ray research who was the first to photograph tracks of mesons (μ, π), a discovery later announced by C. F. Powell. This [M N Saha, J C Bose, J C Ghosh, Snehamoy Dutt, S N Bose, D M Bose, N R Sen, J N Mukherjee, N C Nag] has been obtained by the author(s) from the Wikimedia website, where it is stated to have been released into the public domain. It is included within this book on that basis.

demonstrate the directivity and quasi-optical properties of millimeter waves (Bose 1897; Emerson 1997). We have named the horn antenna and the experiment, BHARAT (Bose Horn Antenna RAdio Telescope), in his honor.

3.3 Instrument

Since HI is a relatively bright and extended source, that is, the source fills the antenna beam, even a small antenna can detect it. A directional antenna is preferred as it allows localization of the features of the HI line to a region in the sky, which helps determine the rotation curve and the Galactic structure. A directional antenna also improves the signal-to-noise ratio to some extent. Several possible antennas can be used, for example, dish antenna, helix antenna or a Yagi antenna. We have used a horn antenna for this experiment. The antenna subsection discusses the antenna properties.

The radio frequency electronic components required for this experiment are challenging to design, assemble and test, and often require special test equipments and skills for fabrication. Fortunately, the advent of satellite based entertainment systems and mobile communication systems together with mass produced computers and associated technologies have made many required components available off-the-shelf at low cost. Hence, off-the-shelf pre-assembled electronics are preferable. Appendix C lists the sources for acquiring the electronics. We look at some antenna and electronics fundamentals below.

3.3.1 Antenna

An antenna converts the electromagnetic fields into electrical signals (current). These signals can then be amplified, filtered and digitized for further analysis. The electromagnetic fields illuminate the antenna aperture and undergo diffraction. The diffraction pattern gives the antenna a specific radiation pattern. These radiation patterns can be estimated as well as measured. The radiation pattern gives us information on the sensitivity of the antenna in various directions. In Figure 3.2, we can see the radiation pattern has a main lobe, side lobes and a back lobe. The main lobe has the maximum sensitivity and is in the direction perpendicular to the antenna aperture. Intuitively, this means that the aperture collects maximum radiation from the direction perpendicular to it. The side lobes and the back lobe indicate that the antenna has some sensitivity to radiation coming from directions other than the antenna pointing. If there is a strong radiator in the direction of these lobes, it will add to the desired measurement introducing noise or interference. Since we are interested only in the radiation coming from the source in the main lobe of the antenna, it is desirable to reduce the side lobes. This can be achieved as follows.

For aperture antennas, the radiation pattern is the Fourier transform of the electric field distribution on the antenna aperture. For a uniformly illuminated circular aperture, the radiation pattern will be an airy pattern with a main lobe and side lobes. If the illumination is tapered near the edge of the aperture, the side lobes will reduce and the main lobe will become broader. We can trade the angular width of the main lobe for a reduction in the antenna side lobes.

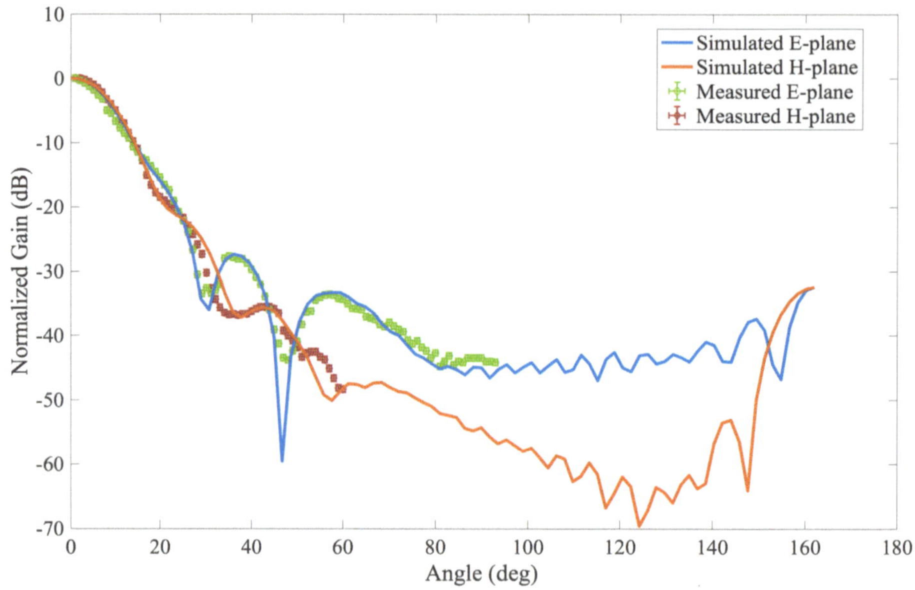

Figure 3.2. The radiation pattern for the dual-mode horn antenna. Both E and H plane measurement match very well with the FEKO (a commercial electromagnetic solver) simulated beams. The nulls are not measured very well due to sensitivity limitations below -40 dB. The beam width is 16.5°, the gain is 20.6 dBi and the worst side-lobe level is <27 dB. This gives excellent beam characteristics for astronomy radiometry.

Since the aperture is no longer uniformly illuminated, we are not utilizing the complete physical aperture. The effective aperture (A_e) is now less than the physical aperture (A).

$$A_e = \eta A \tag{3.1}$$

η is the proportionality constant between 0 and 1, known as aperture efficiency.

The dual-mode horn used in this experiment achieves very low side lobes by tapering the electric field distribution on the aperture. The fundamental mode (TE11) of the circular waveguide is exited in the waveguide. The mode propagates through the waveguide and the horn flare. A step is introduced in the horn flare to generate a second mode (TM11) as seen in Figure 3.3, which also propagates through the flare along with the fundamental mode. The two modes superimpose at the aperture so that the field in the center adds constructively, and the field at the edge adds to reduce the net electric field. This gives a tapered field distribution at the aperture, reducing the side lobes. This also improves the circularity of the beam. The full width half maximum (FWHM) for the dual-mode horn is 16.5°.

Antenna gain (G) is power radiated by the antenna relative to the same power radiated by an isotropic radiator. Since the antenna will radiate different power in different direction, antenna gain is a function of direction. Usually, the maximum gain is quoted for antennas. The gain for the dual-mode horn is 20.6 dBi (dB relative to isotropic radiator). We can get the effective aperture from the antenna gain as follows.

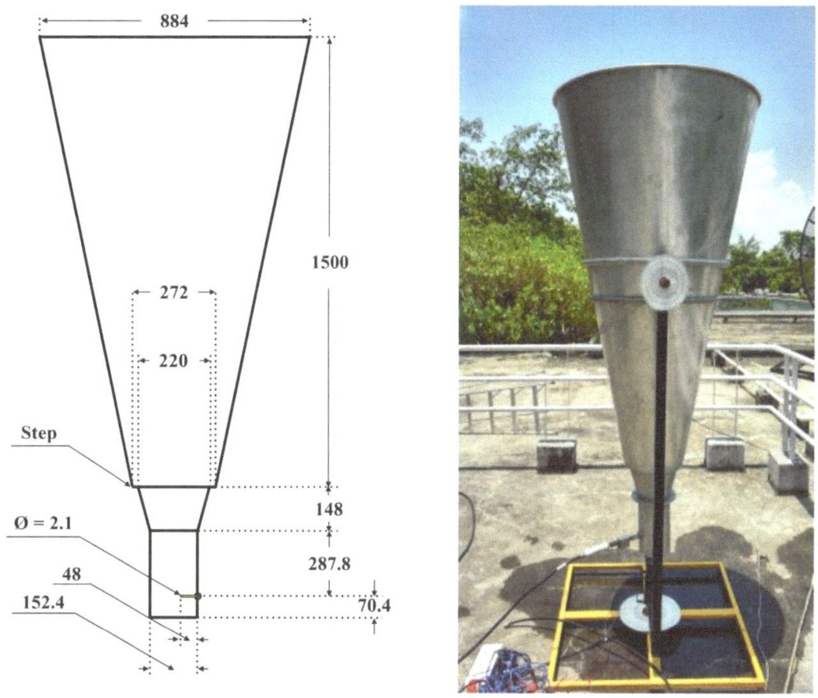

Figure 3.3. The dual-mode horn dimensions (in mm). These are internal dimensions of the horn. Reprinted from Mhaske et al. (2022), with the permission of AIP Publishing.

$$A_e = \frac{\lambda^2 G}{4\pi} = 0.4068 \text{ m}^{-2} \tag{3.2}$$

Using Equations (3.1) and (3.2), we find that the antenna aperture efficiency is 66.28%.

The first component that connects to the antenna is a low-noise amplifier (LNA). The input impedance of the LNA must be matched to the impedance of the antenna port. If this is not the case, then the power coming from the antenna will not be efficiently transferred to LNA and some part of it will be reflected back. In astronomy, sources are usually very weak. Hence, power loss due to impedance mismatch is not desirable. Since the LNA has an impedance of 50 Ohm, the antenna port should be close to this value for a good impedance match. 50 Ohm is one of the industry standards for radio frequency impedance. The input and output impedance are usually mentioned in the electronics datasheet. This parameter should be considered while designing the antenna. The impedance matching of an antenna can be measured by measuring the voltage standing wave ratio (VSWR). This parameter can be measured with a vector network analyzer. It is given in terms of reflection coefficient as

$$VSWR = \frac{1 + |\tau|}{1 - |\tau|} \tag{3.3}$$

where τ is the reflection coefficient.

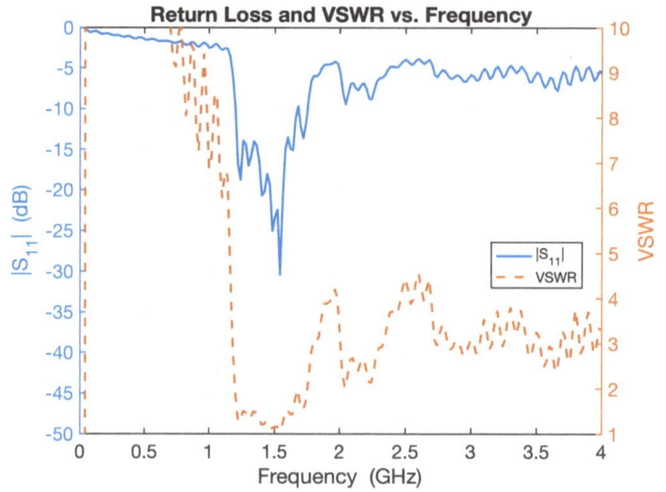

Figure 3.4. The $|S_{11}|$ for the dual-mode horn is better than −20 dB, showing a good impedance match at the antenna port to 50 Ohms. The corresponding VSWR is also plotted. There is good rejection below 1 GHz (good for rejecting mobile bands at ~800–900 MHz) and also above 2 GHz (good for rejecting WiFi signals at 2.4 GHz). The plot also shows the narrow operation bandwidth of the antenna 1.2 GHz to 1.7 GHz. This is sufficient for the 21 cm experiment.

$$\tau = \frac{Z_l - Z_O}{Z_l + Z_O} \tag{3.4}$$

Z_l and Z_O are the load impedance and the reference impedance, respectively. The reference impedance is 50 Ohm in this case. The ideal value of VSWR is 1. In practical cases, it is greater than 1. The closer is the VSWR value to 1, the better is the impedance match. The measured VSWR for the dual-mode horn is 1.11, which implies that the antenna is very well matched to a 50 Ohm load as seen in Figure 3.4.

3.3.2 Electronics

The electronics is divided into the front-end electronics and the back-end electronics. The front-end electronics consists of components such as amplifiers and filters that connect directly to the antenna, whereas the back-end usually consists of additional signal processing electronics. Both of these will be discussed in this section.

3.3.2.1 Front-End Electronics

Since astronomical sources are usually very weak, the first component that connects to the antenna is the LNA. It amplifies the signal while adding very little noise to the signal, which is why it is called a low-noise amplifier. The amplifier used in this experiment has a gain of 37 dB and noise figure of 0.27. The noise figure (NF) is the log scale conversion of the noise factor. The noise factor (F) indicates how much noise the amplifier adds to the input signal.

$$F = 1 + \frac{T_n}{T_0} \tag{3.5}$$

$$\text{NF} = 10\,\log_{10}(F) \tag{3.6}$$

It can be written in terms of the noise temperature T_n and the reference temperature ($T_0 = 290$ K). Using Equations (3.5) and (3.6), we find that the 0.27 dB noise figure corresponds to 18.6 K noise temperature.

We can understand the effect of the LNA with the following equation.

$$T_{\text{eff}} = T_1 + \frac{T_2}{G_1} + \frac{T_3}{G_1 G_2} + \frac{T_3}{G_1 G_2 G_3}\cdots \tag{3.7}$$

Here T_{eff} is the effective noise temperature of the system, T_1 and G_1 are the noise temperature and gain of the first stage of electronics, respectively, T_2 and G_2 are the noise temperature and gain of the second stage of electronics and so on. We can see that the noise of the first stage adds directly to the effective noise temperature of the receiver, whereas the noise temperatures of the subsequent stages is divided by the gain of the first and subsequent stages. Hence, a high gain and low noise figure are desirable for the first LNA.

The 1420 MHz band is reserved for astronomy and there are no communication channels close to this frequency. But there are strong signals, like mobile communication at 900 MHz (this can be different depending on the country) or WiFi and Bluetooth at 2.4 GHz. LNAs are usually broadband devices. They can also operate linearly over a certain range of input power. If strong signals within the operating range of the LNA enter at the input, they can make the response of LNA nonlinear, thereby distorting the signal. If two or more such strong signals are present at the input of a nonlinear amplifier, they can create harmonics called intermodulation products. Although the strong signal may not be present at 1420 MHz, its harmonics can appear at 1420 MHz and corrupt the HI signal. It could also damage the LNA. It is advised to do a radio frequency survey of the site and to use good quality connectors for amplifiers to avoid strong signals entering the system. The antenna can be sensitive to such interferences. Hence low sidelobes and high directivity are desirable in urban areas for sensitive radio astronomy experiments.

The filter[2,3] is connected after the LNA. It filters out any out-of-band signal that could affect the subsequent components in the chain. These two components form the front-end of the electronic chain, but it is possible to have more than one RF LNA and filter in the chain. Too much gain can saturate the receiver, and too little gain can give a poor signal amplification. It can be a good exercise for students to determine the number and sequence of the components in the RF chain using the equations and discussion in this section. For the current experiment, one LNA and one filter were used in the front-end.

[2] https://www.radioastronomysupplies.com

[3] https://www.minicircuits.com

3.3.2.2 Back-end Electronics

The back-end electronics are connected to the front-end electronics with an RF cable as seen in Figure 3.5. Since we are maintaining 50 Ohm impedance in the system, any low-loss 50 Ohm cable can be connected. Examples are RG58 and LMR400. The back-end consists of the software-defined radio receiver (SDR). The SDR can be connected to a PC directly or via a USB cable. In case a USB cable is used, it should be kept short to avoid the addition of noise over the USB cable. For longer lengths, one can opt for longer active USB cables with RF chokes to maintain low noise. Ideally, the SDR should be connected directly to the computer.

An SDR receiver is a special type of receiver that can be controlled via software. RTL-SDR[4,5] was used in the current experiment. It has an amplifier, a filter, a downconverter and an analog-to-digital converter (ADC). Some parameters like the tuner frequency, amplifier gain, sampling rate, etc., can be controlled with software commands. The input signal is at 1420 MHz. This signal is amplified and down-converted to the baseband. The sampling rate sets the bandwidth of the signal. So the baseband signal ranges from DC to $f_{\text{sampling}} = \frac{1}{\text{sampling\_rate}}$. This is possible due to the in-phase quadrature-phase (IQ) sampling in the SDR. The gain and sampling rate were set to 50 and 2.8 megasamples per second (Msps) for this experiment.

The back-end electronics are kept away from the antenna to keep their noise and interference from entering the antenna or affecting the front-end electronics.

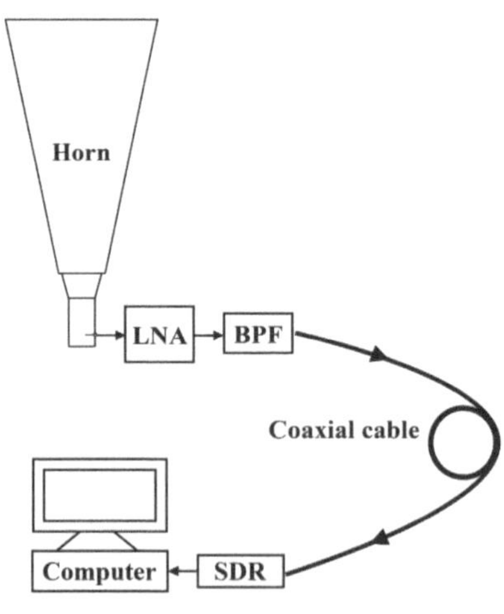

Figure 3.5. The instrument setup: front-end consists of a LNA and a band pass filter while the back-end is the SDR and the computer. The cable connecting the front-end electronics to the back-end receiver is a RF cable, e.g., LMR400 or RG58 with 50 Ohm characteristic impedance. Reprinted from Mhaske et al. (2022), with the permission of AIP Publishing.

[4] https://www.rtl-sdr.com/buy-rtl-sdr-dvb-t-dongles/
[5] https://www.nooelec.com/store/

3.4 Radiometer Theory

It is essential to establish the connection between the antenna temperature and the temperature of the astronomical source for the radio telescope measurements. This would allow us to interpret the power measured in a radio telescope in terms of the intrinsic properties of celestial sources, such as their brightness temperature, giving meaningful astrophysical interpretations. The following argument given by Robert H. Dicke provides this relation (Dicke 1946).

A perfect blackbody absorbs all incident electromagnetic radiation and, when in thermal equilibrium at temperature T, emits radiation with a characteristic spectrum that depends solely on T. The spectral radiance $B_\nu(T)$, the emitted power per unit area, per unit frequency, per unit solid angle, is given by Planck's Law.

$$B_\nu(T) = \frac{2h\nu^3}{c^2}\frac{1}{e^{\frac{h\nu}{kT}} - 1}\ (W\ m^{-2}\ Hz^{-1}\ sr^{-1}) \tag{3.8}$$

where h is Planck's constant, ν is frequency, c is the speed of light, k is the Boltzmann's constant, and T is the absolute temperature in Kelvin. In the radio frequency regime, where $h\nu \ll kT$, the exponential term can be expanded using a Taylor series,

$$e^{\frac{h\nu}{kT}} = 1 + \frac{h\nu}{kT} \tag{3.9}$$

Substituting this in Planck's Law gives the Rayleigh–Jeans Law,

$$B_\nu(T) \approx \frac{2kT\nu^2}{c^2} = \frac{2kT}{\lambda^2} \tag{3.10}$$

In a radio telescope, antenna intercepts incident radiation over an effective area A_e. For an antenna with gain G at wavelength λ the effective area is

$$A_e = \frac{G\lambda^2}{4\pi} \tag{3.11}$$

Assume an antenna connected to a matched resistor and surrounded by a blackbody radiating temperature T. The power per unit bandwidth P_ν received from a source with brightness distribution $B_\nu(\theta, \phi)$ over the antenna beam is,

$$P_\nu = \frac{1}{2}\iint_{\text{source}} A_e(\theta, \phi)B_\nu(\theta, \phi)d\Omega \tag{3.12}$$

The factor 1/2 is to account for the reception of single polarization. Using Equation (3.10), we get

$$P_\nu = \frac{kT}{\lambda^2}\iint_{4\pi} A_e(\theta, \phi)d\Omega \tag{3.13}$$

For a well-designed antenna, integrating over all directions for gain in Equation (3.11) will give

$$\iint_{4\pi} A_e(\theta, \phi) d\Omega = \lambda^2 \tag{3.14}$$

This leads to Equation (3.12) becoming

$$P_{\nu,\text{antenna}} = kT \tag{3.15}$$

In 1928, Nyquist (Nyquist 1928) demonstrated that a resistor at temperature T produces thermal noise with available power per unit frequency bandwidth,

$$P_{\nu,\text{resistor}} = kT \tag{3.16}$$

This is the spectral power generated in the resistor in a thermal bath of temperature T. Now consider our antenna setup, which is terminated with a matched resistor and is placed in a uniform blackbody radiation bath. By the second law of thermodynamics, no net energy flow occurs between components at the same temperature. If the resistor delivers kT power per unit bandwidth, the antenna must deliver kT power per unit bandwidth to the resistor. Therefore, measuring the resistor temperature gives us the antenna temperature, which in turn gives us the intrinsic measurement of the source temperature. The thermodynamic argument ensures that this representation is physically grounded.

3.5 Calibration

Calibrating an instrument involves conversion of the measurements using the instrument into physical or absolute units for comparison with the outcome of other similar experiments. The setup described in the previous section is a radiometer, i.e., a device used to measure temperature in the radio band. We measure the power at the output of the receiver. In an ideal case, if the antenna is looking at a source of temperature T and we measure power P at the antenna load (perfectly matched), then we can estimate the source power using $P = kT$, where k is the Boltzmann constant. In reality, the total temperature of the system T_{sys} measured by the receiver is not just the temperature of the source T_{source} but also has contributions from the receiver T_{receiver} and other factors like the side lobe and the back lobe temperature. Factors like ambient temperature, power supply voltage fluctuations and amplifier gain fluctuations can also affect T_{sys}:

$$T_{\text{total}} = T_{\text{source}} + T_{\text{receiver}} + T_{\text{sidelobes}} + T_{\text{backlobe}} + T_{\text{sky}} \cdots \tag{3.17}$$

We will try to understand the implication of T_{sys} and calibrate the data in this section. Since the HI spectrum has information on the dynamics of the source, the x-axis can be converted from frequency to velocity to get this information.

3.5.1 Absolute Calibration

In the radiometer, the signal from the source passes through various components in the receiver chain. There can be a multiplication factor and an offset added to the signal by the system. These factors need to be determined. Effectively, this means we need to determine the contribution of the receiver T_{receiver} to the source signal.

To determine the two unknowns, i.e., scaling and offset, of the receiver, we need to form two linear equations. If we measure the power from two sources with known temperatures, we can get these factors and use them to convert the power of an unknown source into temperature. For the purpose of this experiment, we consider the ground and the sky as the two reference sources for calibration. It is apparent from the measurements that these are valid sources for our conversion. We use a method derived from the On-Off method or the Y-Factor method to calibrate the instrument.[6]

When the antenna points to a source, the measured power has contributions from the source, the receiver, and other factors mentioned in Equation (3.17). Due to the low side lobes and back lobes of the dual-mode horn, we can assume that the temperature detected by the antenna comes only from the source in the main beam of the antenna. Therefore, we can simplify Equation (3.17) to have only two terms on the right-hand side. We assume a simple model for the system as follows,

$$P = G(T_{\text{receiver}} + T_{\text{ant}}) \tag{3.18}$$

Here, P is the power measured at the end of the receiver, G is the system gain, T_{ant} is the antenna temperature and by extension, the temperature of the source in the antenna beam (assuming negligible losses in the antenna). We need to determine G and T_{receiver} to convert a source power measurement into its brightness temperature. When the antenna points to the ground, we get

$$P_{\text{ground}} = G(T_{\text{receiver}} + T_{\text{ground}}) \tag{3.19}$$

and when the antenna points to the blank sky close to the zenith but away from the Galactic plane, we get

$$P_{\text{sky}} = G(T_{\text{receiver}} + T_{\text{sky}}) \tag{3.20}$$

These can be seen in Figure 3.6. Taking the ratio of P_{ground} and P_{sky}

$$\frac{P_{\text{ground}}}{P_{\text{sky}}} = \frac{T_{\text{ground}} + T_{\text{receiver}}}{T_{\text{sky}} + T_{\text{receiver}}} \tag{3.21}$$

we can determine T_{receiver} as

$$T_{\text{receiver}} = \frac{T_{\text{sky}}\frac{P_{\text{ground}}}{P_{\text{sky}}} - T_{\text{ground}}}{1 - \frac{P_{\text{ground}}}{P_{\text{sky}}}} \tag{3.22}$$

where P_{ground} is the power measured when the antenna is pointed to the ground, P_{sky} is the power measured when the antenna is pointed to the sky, T_{ground} is the ground temperature (assumed as 290 K) and T_{sky} is the sky temperature assumed as 7.2 K at the zenith (discussed in the next subsection).

[6] https://eventhorizontelescope.org/sites/g/files/omnuum3116/files/eht/files/EHT_memo_Issaoun_2017-CE-02.pdf

Substituting Equation (3.22) in (3.21) we get G as

$$G = \frac{P_{\text{ground}} - P_{\text{sky}}}{T_{\text{ground}} - T_{\text{sky}}} \tag{3.23}$$

G has units of Watts/Kelvin, implying that this is a conversion gain from power to temperature. Equations (3.22) and (3.23) are the calibration equations. We can point the antenna to the HI source to get

$$P_{\text{source}} = G(T_{\text{receiver}} + T_{\text{source}}) \tag{3.24}$$

Solving this for T_{source} gives

$$T_{\text{source}} = \frac{P_{\text{source}}}{G} - T_{\text{receiver}} \tag{3.25}$$

Since T_{source} measurement inherently includes T_{sky}, we can subtract T_{sky} to get the final T_{source} as

$$T_{\text{source}} = \frac{P_{\text{source}}}{G} - T_{\text{receiver}} - T_{\text{sky}} \tag{3.26}$$

A caveat of this method is that HI is present in all directions and appears as a small peak in all T_{sky} measurements, as seen in Figures 3.6 and 3.7. Since T_{sky} affects both the calibration Equations (3.22) and (3.23), we need to correct T_{sky} before using it in calibration. One way to do this is to use frequency switching for sky measurement, i.e., use a nearby frequency to measure sky temperature instead of 1420.405 MHz. However, if the front-end filter has a frequency-dependent response, this may not give a linear correction to the observation.

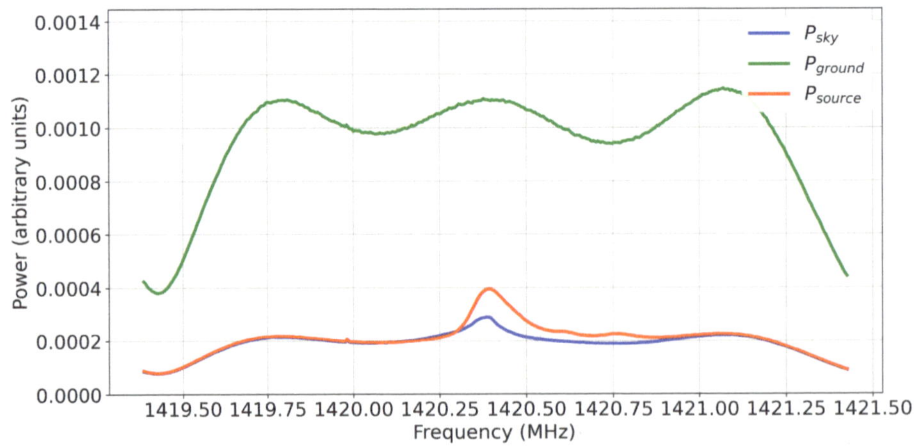

Figure 3.6. Spectrum of the three antenna measurements, P_{ground}, P_{sky} and P_{source}. The shape of the spectrum is mainly due to the narrow band filter present in the receiver. This shape gets removed by following the calibration procedure.

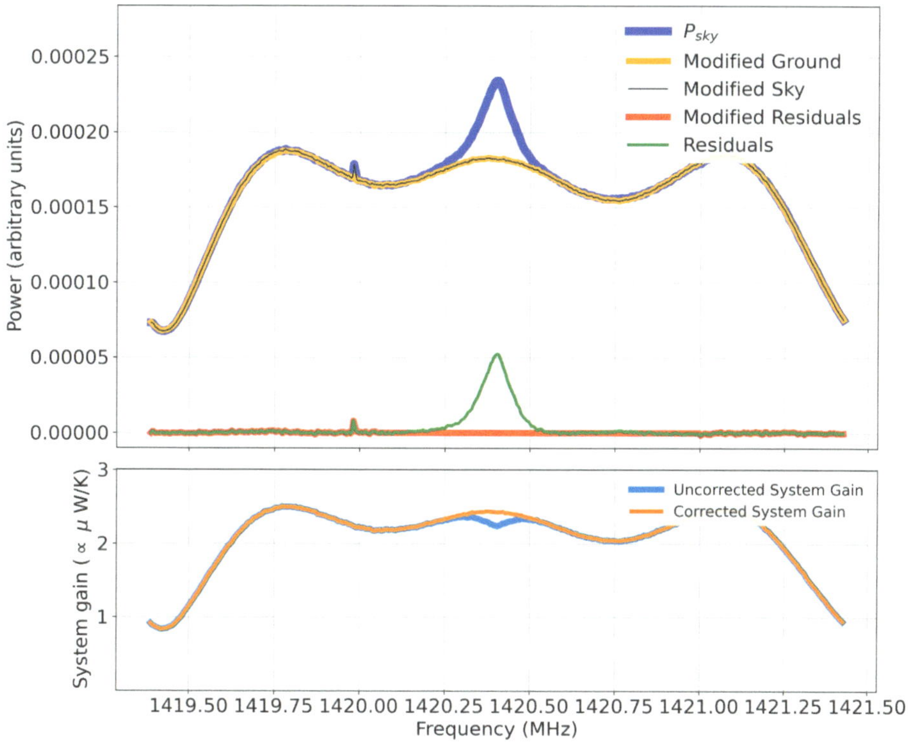

Figure 3.7. Correction to P_{sky} using P_{ground} in portion of P_{sky} where the HI line is present. The top panel shows modified ground after scaling and modified sky measurement after substitution. The residuals before and after are also plotted. The bottom panel shows system gain before and after P_{sky} correction.

Although this is a viable solution, we use a different technique here. If the system is linear, then different blackbody source temperature measurements are scaled versions of each other. We can scale the ground measurement to match the sky measurement and replace the middle portion of the sky measurement with the modified ground measurement. We are treating the ground measurement as a model for the system. We do not replace the entire sky measurement with the ground measurement because the other portion of the sky measurement is useful. If there are internal interference peaks in the receiver, the calibration method will remove them from the final result. An example is the small peak at 1420 MHz in Figure 3.7. Once we have the modified sky measurement, we can proceed with the calibration process described earlier in this section. We remove the offset in the source temperature to get the brightness temperature of the HI line as seen in Figure 3.8.

As is apparent from this figure, the calibration method is effective in removing almost all features present in the raw measurement data in Figure 3.6.

3.5.2 Atmosphere and Other Considerations

For ground measurement, the antenna can be pointed to a nearby wall or ground so that it fills the antenna beam. During pointing, the antenna should not be too close

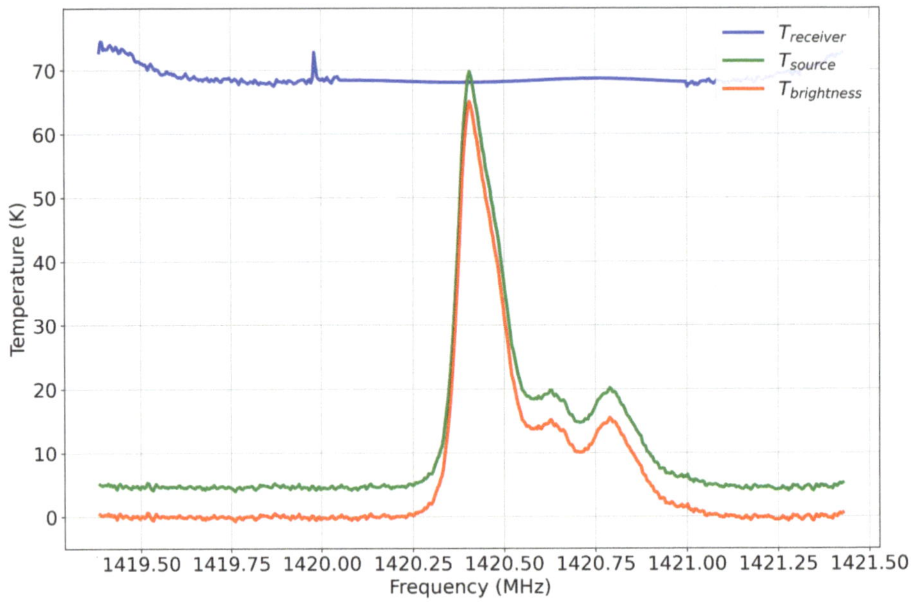

Figure 3.8. Calibrated brightness temperature after removing the offset in T_{source}. The smooth central part of T_{receiver} is due to the substitution of modified ground to remove the HI line in P_{sky}

to the wall and should not be facing the wall directly. This could create standing waves between the antenna and the wall and could saturate the amplifier. It is recommended to point the antenna at a slanting angle. Ideally, a microwave absorber should be used to avoid reflections and standing waves. To be a bit more precise, the physical temperature of the wall or ground can be measured and used as T_{ground}.

For sky measurement, the zenith is the most preferable position. The sky temperature has three components, the cosmic microwave background temperature ($T_{\text{cmb}} = 2.7$ K), the atmospheric contribution ($T_{\text{atm}} \approx 2.9$ K), and the background radio continuum temperature, $T_{\text{cont}} \approx 1.6$ K (Delahaye et al. 2002).

$$\begin{aligned} T_{\text{sky}} &= T_{\text{cmb}} + T_{\text{atm}} + T_{\text{cont}} \\ &= 7.2 \text{ K} (\text{at zenith 20 deg away from Galactic plane}) \end{aligned} \tag{3.27}$$

The background radio temperature ($T_{\text{cmb}} + T_{\text{cont}}$) will be present in all directions. For a better sky temperature estimate, one can use the 1420 MHz continuum survey (Reich 1982; Reich & Reich 1986; Testori et al. 2001; Reich et al. 2001) available online.[7]

The atmospheric contribution is important. The atmosphere is at a physical temperature. Hence, it radiates as well as absorbs its own radiation. The radiation from the upper layers of the atmosphere is strongly attenuated due to absorption in the lower layers. Hence, the layers closer to the ground are the most significant

[7] https://lambda.gsfc.nasa.gov/product/foreground/fg_stockert_villa_get.html

contributors to temperature along the antenna line of sight. This radiative transfer can be summarized in the following equation (Wilson et al. 2013; ITU 2019):

$$T_{\text{atm}} = 290(1 - e^{\frac{-\tau}{\sin(\text{El})}}) \tag{3.28}$$

T_{atm} is the atmosphere's contribution to the antenna temperature, El is the elevation angle and τ is the atmospheric opacity along the line of sight (dimensionless). Opacity is a measure of how much the atmosphere absorbs and emits. It depends on frequency, pressure, temperature, composition, and the path length through the atmosphere. The value of τ is very low at 1.4 GHz, about 0.01 (ITU 2019; Wilson et al. 2013). For elevation angle of 90, i.e., at zenith, and since $\tau \ll 1$, we can use the Taylor approximation of the exponential to get

$$T_{\text{atm}} = 290\tau = 2.9 \text{ K} \tag{3.29}$$

This gives us the contribution of the atmosphere to the antenna temperature. As noted earlier, sky measurement should be taken near zenith but away from the galactic plane since the galactic plane has a higher value of T_{cont}.

For observations near the horizon (low elevation), the measurements will have higher T_{atm} (Equation (3.28)) and contributions from the ground as well. Since the antenna beam is large, for low elevations, some portion of the beam will be pointing to the ground. The antenna will measure a larger source temperature offset, and the brightness temperature measurement will deviate from expected values at these elevations. The measured $T_{\text{brightness}}$ is lower, and the offset is higher than expected for low elevations due to the fact that the beam is not completely filled with the HI source but is partly blocked by the ground. Structures like buildings and trees could also have this effect if they partially block the beam.

It is recommended that the calibration for each source be done if possible or at least within a reasonable time period (15–30 min). This period depends on the sensitivity requirements for the source brightness temperature measurement. The calibration changes along with the change in ambient temperature as T_{receiver} depends on ambient temperature. If there is a long delay between the source measurement and the calibration, there will be errors in $T_{\text{brightness}}$ and the baseline of the plot will not be flat. The calibration should be done at a much faster rate to avoid the effects of gain fluctuations in the system.

3.6 Velocity Correction

Once the Y-axis (temperature) of the plot is calibrated, the X-axis needs to be converted from frequency to velocity. Frequency can be converted to velocity using the Doppler Equation (3.30).

$$V_{\text{LSR}} = c\left(1 - \frac{f}{f_0}\right) + V_{\text{corr}} \tag{3.30}$$

The first term on the right-hand side of Equation (3.30) is the Doppler effect, which converts the measured frequency (f) to velocity with the reference frequency as the rest frequency ($f_0 = 1420.405751 \quad MHz$) of the 21 cm HI line. This conversion gives the velocity of the source relative to the observer. The second term, $V_{\rm corr}$, is the correction term, which removes the effects of the motion of the observer with respect to the local group of stars. The local motion of the observer has three major components, i.e., the rotation of Earth, the revolution of Earth around the Sun and the peculiar motion of the Sun with respect to the mean velocity of the solar neighborhood. The $V_{\rm corr}$ depends on the direction of the source in the sky, the position of the observer on Earth's surface and the observation date and time. The velocity we get after the correction is the velocity of the source with respect to the local standard of rest[8] (LSR). $V_{\rm LSR}$ is now independent of the observer and can be used for comparisons (Figure 3.9).

The accuracy of the frequency measurement depends on the oscillator used inside the RTL-SDR dongle. The oscillator frequency can drift over time due to temperature variation. It is recommended that the SDR dongle be used with a temperature-compensated oscillator (TCXO) for best results. These dongles will have negligible drifts during the experiment. There could be frequency offsets in the oscillator. These offsets are usually minimal. The other way to correct these effects is to compare the plot with a plot obtained from astronomical surveys, where the velocity conversion takes into account several effects. One such comparison shown in Figure 3.10 is discussed in Chapter 4.

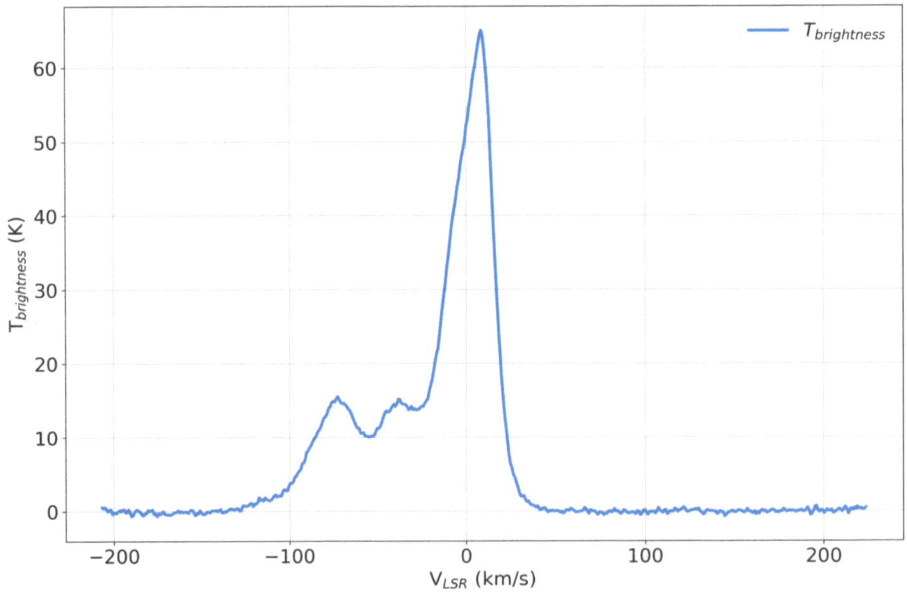

Figure 3.9. The $V_{\rm LSR}$ vs. $T_{\rm brightness}$ after correcting for local motion effects.

[8] The local standard of rest (LSR) is defined as a point in space that is moving on a perfectly circular orbit around the Galactic Centre at the solar position and is usually calculated using the mean of velocities of stars in the solar neighborhood (Huang et al. 2015).

3.7 Antenna Pointing

One could use calibrated angle markings and a pointer for azimuth and elevation angles to point the antenna. The elevation angle goes from 0° at horizon to 90° at zenith. The azimuth angle goes from 0° in the North direction and increases clockwise to 360° in the horizon plane. Modern smartphones have accelerometers and magnetometers that can give pointing angles. The accelerometer can measure the direction of local acceleration due to gravity, which can be converted into an elevation angle. The magnetometer measures the orientation of the sensor with respect to the Earth's magnetic field. This measurement can be converted to the azimuth angle measurement. There are several mobile phone apps that can give these values using the sensors available in the phone.[9,10] Placing the mobile phone flat against the back of the horn can give the azimuth and elevation coordinates of the pointing direction of the horn. It is recommended to measure azimuth angle by keeping the elevation angle 0° as the apps may not compensate for magnetometer tilts. Orientation of the phone depends on the application used.

Stellarium[11] can be used to determine a source's altitude and azimuth angles. Since these angles are in the local frame of the observer, Stellarium uses the location and current date–time to determine the position in the sky as visible from the antenna location.

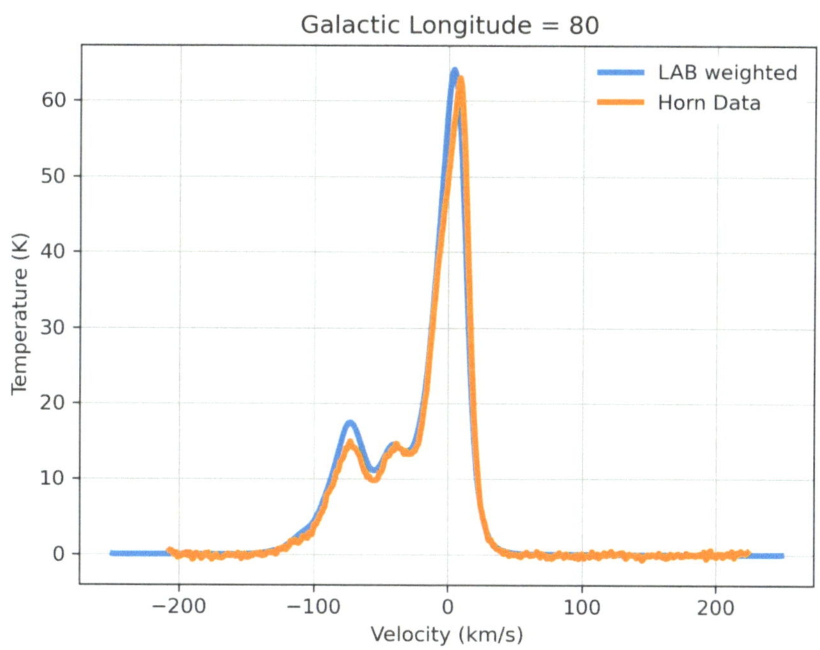

Figure 3.10. Comparison of the calibrated data with weighted LAB survey for galactic longitude 80.

[9] https://play.google.com/store/apps/details?id=com.esys.satfinder&hl=en&gl=US
[10] https://play.google.com/store/search?q=sky%20portal&c=apps&hl=en&gl=US
[11] https://stellarium.org

3.8 Data Fitting and Sensitivity

Data fitting gives students an introduction to data analysis. Here, we explore the concepts of data fitting and residuals and connect them to the physical parameters of the system. We fit multiple Gaussians to the line profile. The sum of these Gaussians is the model for the observations. If we subtract the model from the data, we get the residuals. If the residuals are noise-like, then the model fits the data well. Any non-random features in residuals tell us about features not accounted for by the model. One can use an iterative process to fit multiple Gaussians to the data. We look at the residuals and keep adding Gaussians until the residuals are noise-like (i.e., distributed with a normal distribution), as shown in Figure 3.11. For physical relevance, we can set the maximum number of Gaussians to avoid overfitting. The residuals do not always appear noise-like with zero mean, suggesting the non-linearity in the calibration process. The instrument is linear within a certain range of input and output power. As the sensitivity of the measurement increases, one needs to account for several factors like calibration time, impedance matching, amplifier nonlinearity, etc.

The sensitivity of a radiometer is given by the root-mean-square (RMS) noise fluctuation in the measured antenna temperature $\Delta T_{\rm rms}$, as described by the ideal radiometer equation.

$$\Delta T_{\rm rms} = \frac{T_{\rm sys}}{\sqrt{\Delta f \Delta t}} \tag{3.31}$$

Δf is the width of a channel (or frequency step in the spectrum), which is 4 kHz, and Δt is the integration time, which is 60 s in the current measurement setup. This gives $\Delta T_{\rm rms} = 0.163$ K. Since the sky, ground and HI are extended sources, i.e., they fill the antenna beam, $\Delta T_{\rm rms}$ is the standard deviation of the residuals after fitting. If we look at the standard deviation of the points where the HI line is absent, we get a value of $\Delta T_{\rm rms}$ measurement very close to the ideal value. However, after fitting, the standard deviation of all points in residuals is around 0.27 K. This higher value can be attributed to the fact that fitting is not always exact. In addition, $T_{\rm sys}$ is higher where the HI line is present, increasing the total $T_{\rm rms}$. Along with these factors, system nonlinearities also play a part in increasing $T_{\rm rms}$. Ideally, $T_{\rm rms}$ is the RMS fluctuation of an individual point in the plot. Since this would require several observations in the same conditions for the same point, we have assumed that all points in the plot belong to the same noise distribution. This assumption may not always be true, as the system response and nonlinearities in the system could depend on frequency, but it is valid for the current setup.

3.9 Conclusion

The BHARAT setup is an easy-to-assemble radio telescope with off-the-shelf components. In this chapter, we illustrated various aspects of the experiment. We saw how the setup detects the 21 cm HI hydrogen line. Calibration derived from radiometer theory provides the brightness temperature of the line. We can achieve a

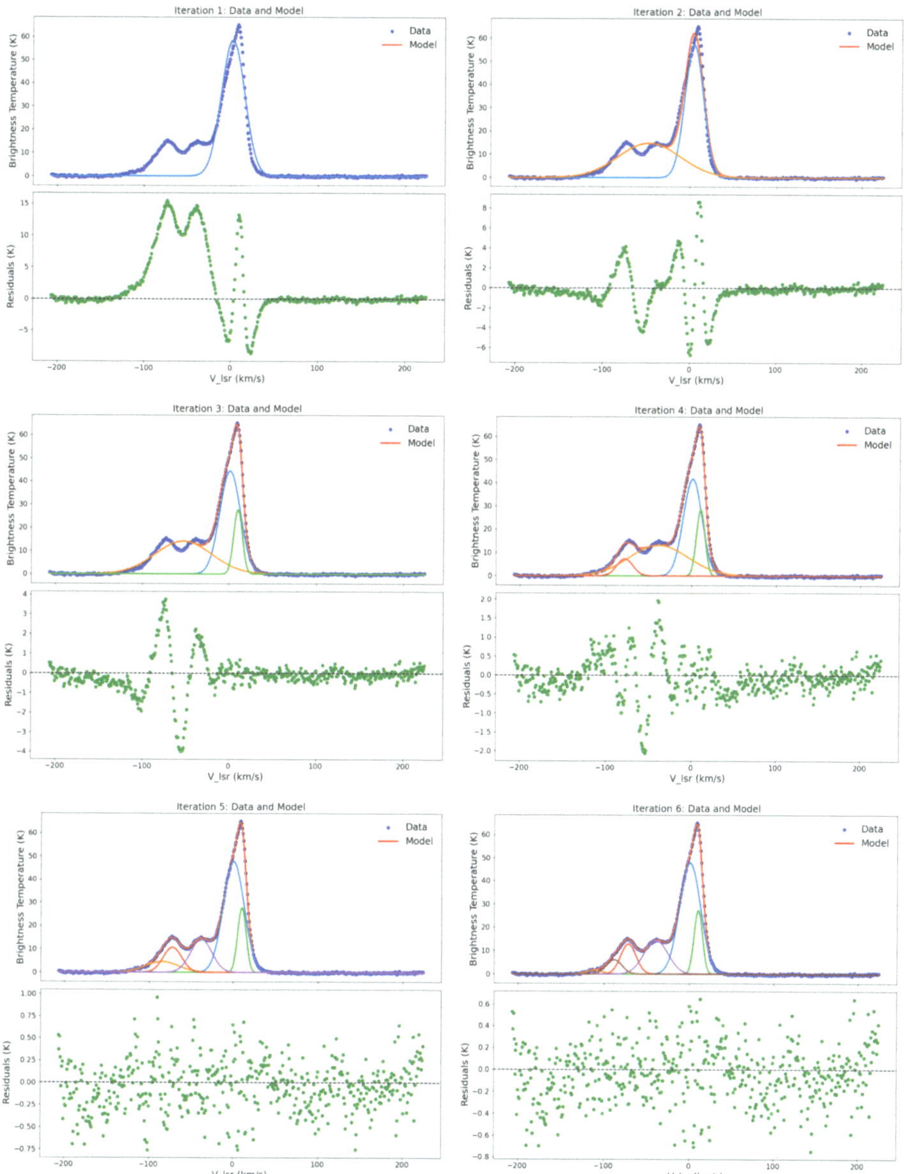

Figure 3.11. Data fitting.

high-sensitivity measurement of the HI line. The astrophysical significance and results derived from these observations are discussed in Chapter 4.

3.10 Appendix A: All Data

All calibrated HI line profiles with the corresponding Gaussian components and the model are plotted in Figures 3.12–3.14.

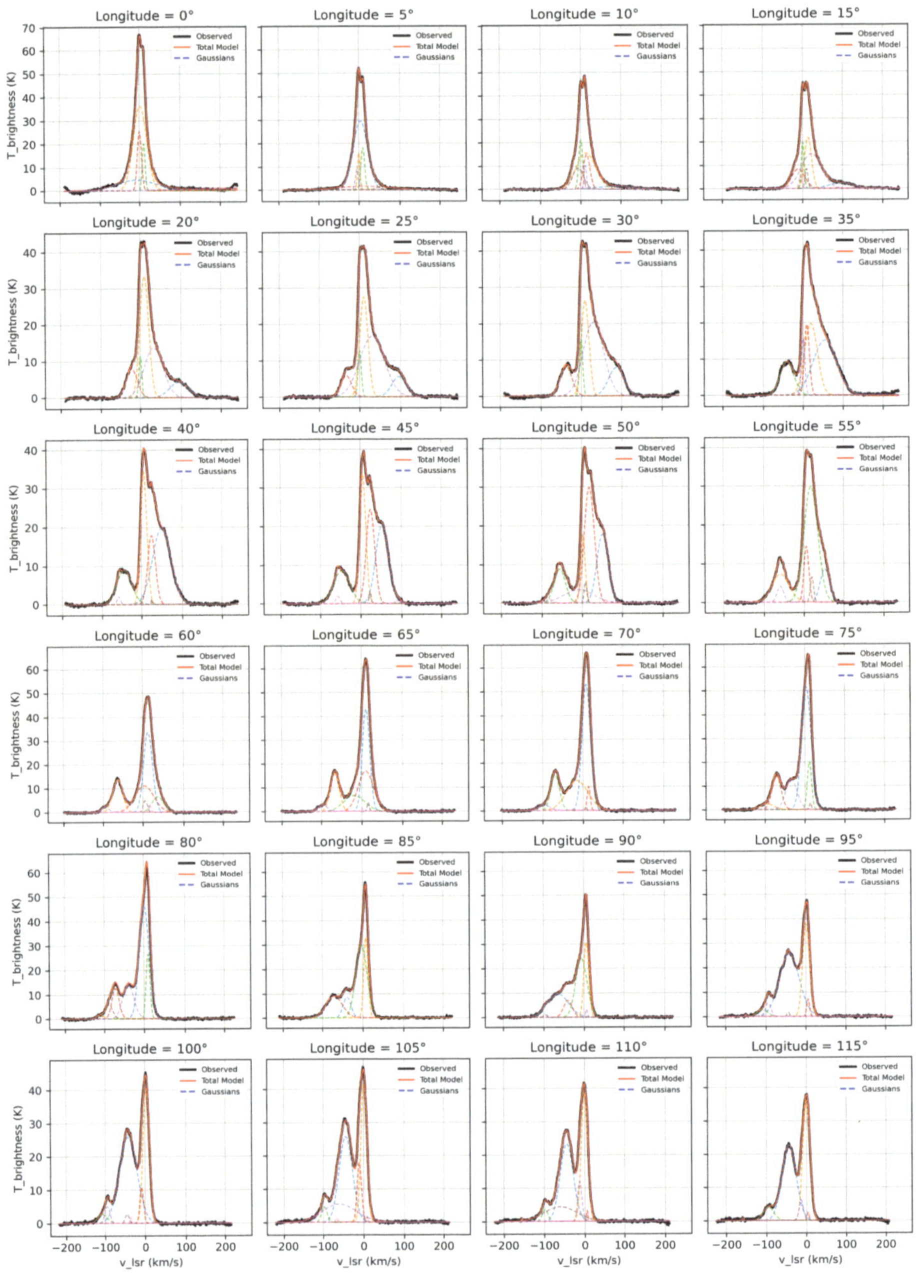

Figure 3.12. All calibrated HI line profiles, 1 of 3.

3.11 Appendix B: Tutorials

Jupyter notebook tutorials are provided in the supplementary material. Following are the tutorials for this chapter.

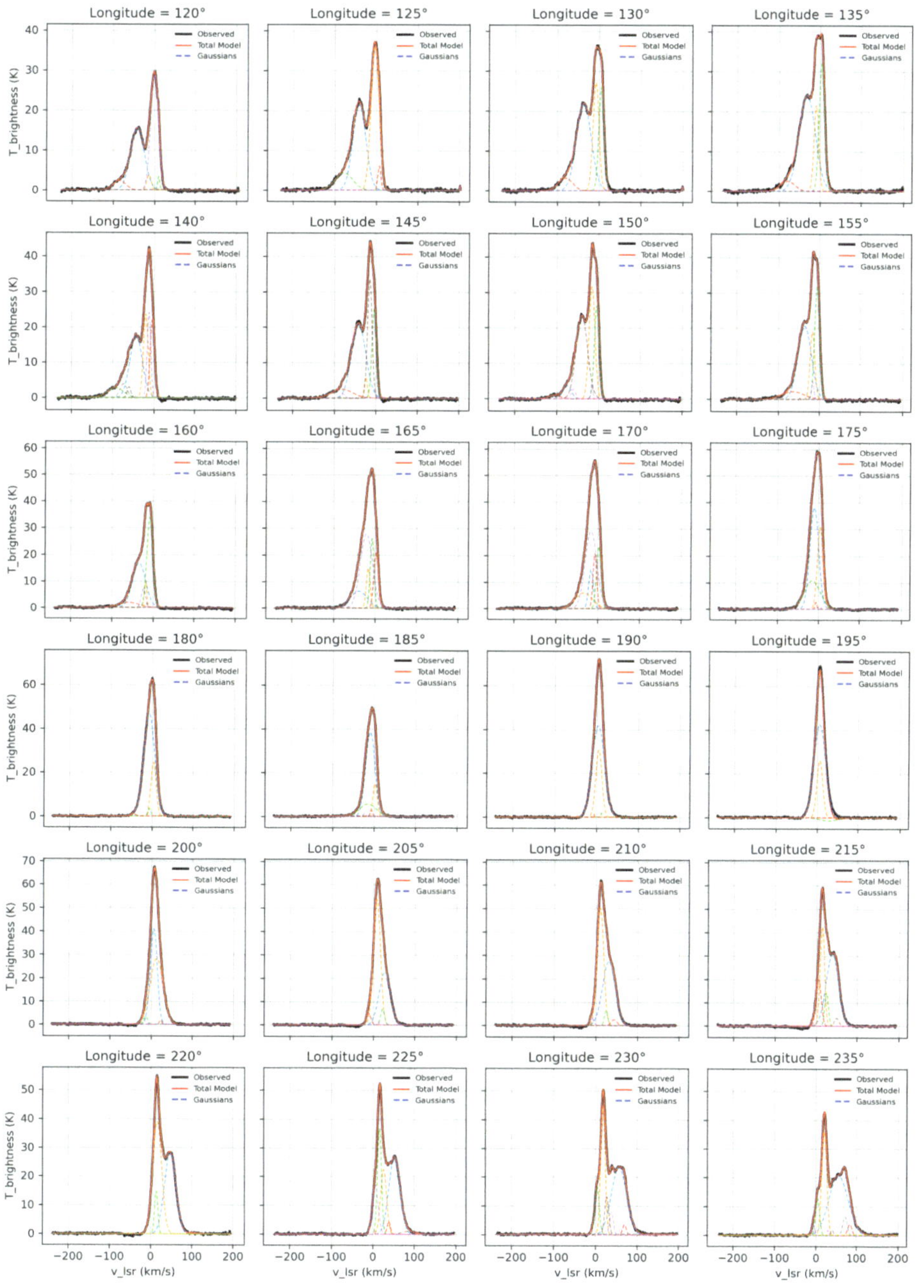

Figure 3.13. All calibrated HI line profiles, 2 of 3

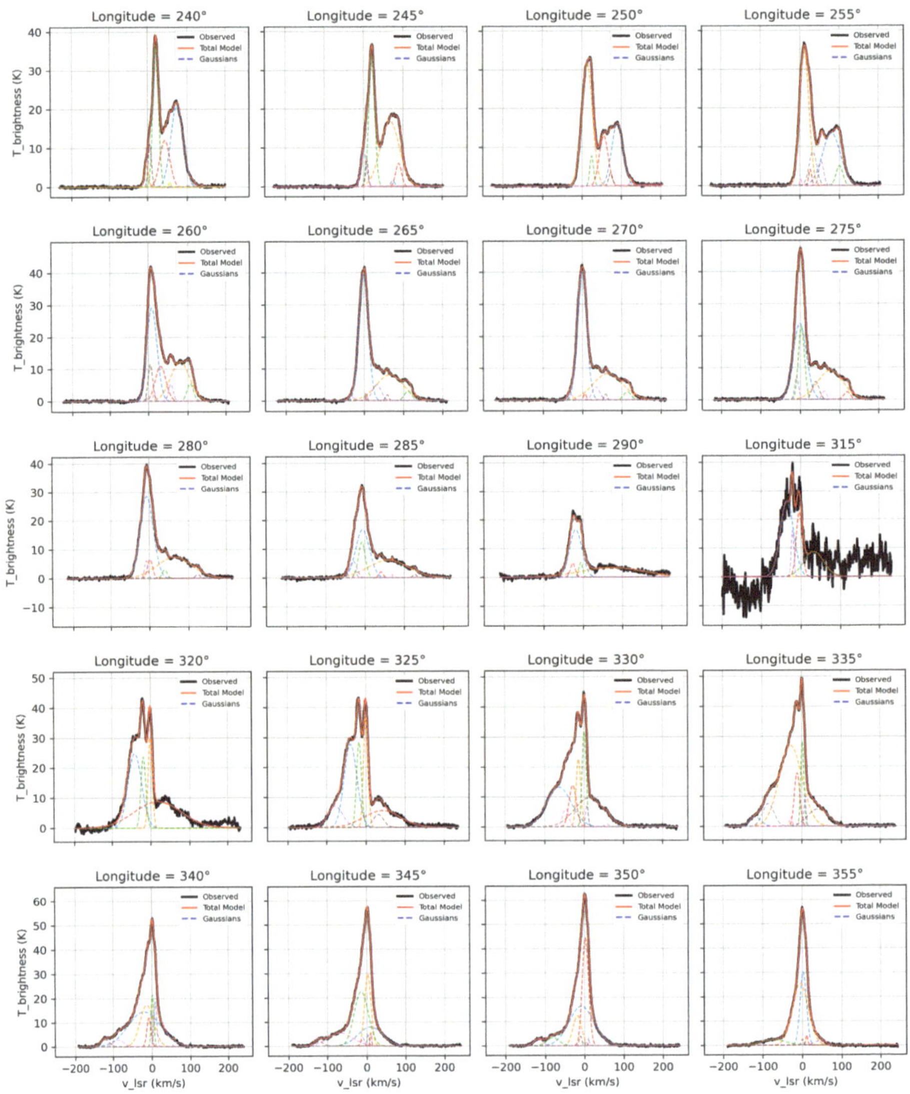

Figure 3.14. All calibrated HI line profiles, 3 of 3.

- Data recording.
- Calibration and Velocity Conversion.
- Data Fitting.

The latest versions of the tutorials can be found in: https://github.com/Ashish-Mhaske/BHARAT_Tutorials.

3.12 Appendix C: Pyramidal Horn

The following is a design of a compact pyramidal horn for easier assembly compared to the dual-mode conical horn. Figures 3.15–3.17 show the dimensions, radiation

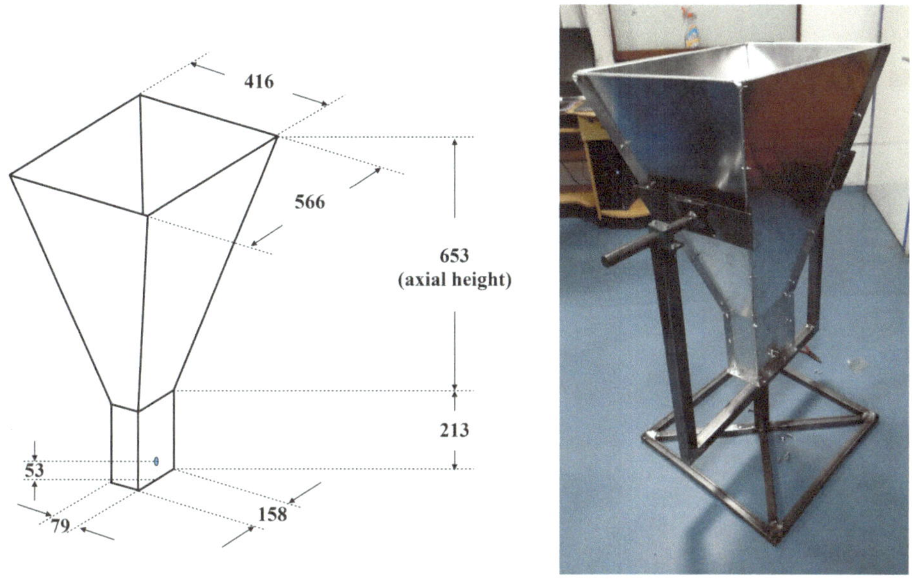

Figure 3.15. Dimensions and the assembled Pyramidal horn. The antenna is mounted on a simple Alt-Az mount with locks for fixing the antenna pointing angles during observations. Reprinted from Mhaske et al. (2022), with the permission of AIP Publishing.

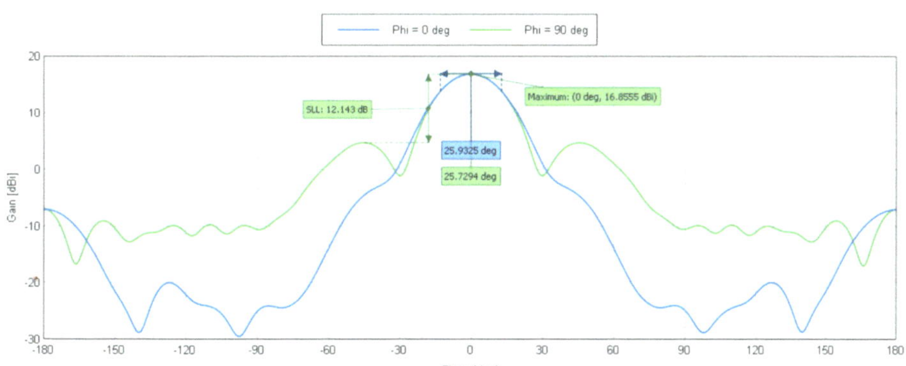

Figure 3.16. The beam simulation of the pyramidal horn as carried out in FEKO. The beam width is about 26° deg, gain is 16.85 dBi and the worst side-lobe level is 12.1 dB.

pattern, and reflection coefficient measurement for the Pyramidal horn. The noise in this setup is usually higher ($\approx$ twice that of the dual-mode) horn setup due to the comparatively poor sidelobes. But this antenna is easier to fabricate and, since HI is a bright source, sufficient for the detection of the 21 cm line.

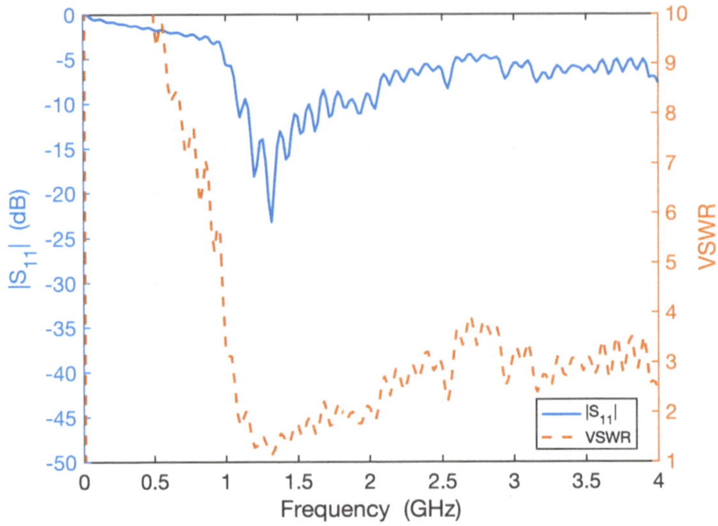

Figure 3.17. S11 measurement shows a good impedance match of the pyramidal horn at 1420 MHz with $|S_{11}| < 16 \quad dB$.

References

Arul Pandian, B., Ganesh, L., Inbanathan, S. S. R., et al. 2022, arXiv:2202.11039

Bandura, K., Prestage, R., & Sanghavi, P. 2018, 2018 2nd URSI Atlantic Radio Science Meeting (AT-RASC) (Piscataway, NJ: IEEE) 1

Berkhout, L. M., Beardsley, A. P., Jacobs, D. C., et al. 2023, PhyEd, 59, 015020

Bose, J. C. 1897, RSPS, 60, 433

Delahaye, J. Y., Golé, P., & Waldteufel, P. 2002, RaSc, 37, 11

Dicke, R. H. 1946, RScI, 17, 268

Dicke, R. H., Peebles, P. J. E., Roll, P. G., & Wilkinson, D. T. 1965, ApJ, 142, 414

Emerson, D. T. 1997, ITMTT, 45, 2267

Ewen, H. I., & Purcell, E. M. 1951, Natur, 168, 356

Findlay, J. W. 1966, ARA&A, 4, 77

Fung, L. W. H., Lau, A. W. K., Chan, K. H., & Shing, M. T. 2023, arXiv:2309.15163

Huang, Y., Liu, X.-W., & Yuan, H.-B. 2015, MNRAS, 449, 162

Jansky, K. G. 1933, Proceedings of the Institute of Radio Engineers, 21, 1387

Leech, M. 2017, A 21cm Radio Telescope for the Cost-Conscious https://www.ccera.ca/papers/a-21cm-radio-telescope-for-the-cost-conscious/

Mhaske, A. A., Bagchi, J., Joshi, B. C., Jacob, J., & Paul, K. T. 2022, AmJPh, 90, 948

Nyquist, H. 1928, PhRv, 32, 110

ITU, 2019, Attenuation by Atmospheric Gases and Related Effects P.676-12, *ITU*

Patel, N. A., Patel, R. N., Kimberk, R. S., et al. 2014, AAS Meeting, 224, 415.01

Penzias, A. A., & Wilson, R. W. 1965, ApJ, 142, 419

Reber, G. 1944, ApJ, 100, 279

Reich, P., & Reich, W. 1986, A&AS, 63, 205

Reich, P., Testori, J. C., & Reich, W. 2001, A&A, 376, 861

Reich, W. 1982, A&AS, 48, 219

Rogers, A. E. E. 2012, SRT Memo Series https://www.haystack.mit.edu/haystack-memo-series/srt-memos/

Testori, J. C., Reich, P., Bava, J. A., et al. 2001, A&A, 368, 1123

Wilson, T. L., Rohlfs, K., & Hüttemeister, S. 2013, Tools of Radio Astronomy (New York: Springer)

Radio Telescope Instrumentation for Teaching

Timothy Dolch

Chapter 4

BHARAT Setup for HI Pedagogy: HI Astrophysics

Ashish A. Mhaske, Bhal Chandra Joshi and Dhruba Saikia

We present the Bose Horn Antenna Radio Telescope (BHARAT; Mhaske et al. 2022), a low-cost and low-complexity setup that is being used in colleges and universities for the 21 cm hydrogen experiment. We opt for off-the-shelf components and Python code for easy assembly and operation. Chapter 3 describes the instrument, observations and calibration of the data, while this chapter describes the astrophysics derived from these observations. Tutorials are provided for data collection and data analysis for all the topics discussed in both chapters.

4.1 Introduction

The study of neutral hydrogen (HI) through its 21 cm hyperfine transition has been instrumental in shaping our understanding of the Milky Way's structure and dynamics. Since its first detection by Ewen and Purcell in 1951, HI emission has enabled astronomers to map the distribution and motion of interstellar gas across the Galaxy. Modern all-sky surveys, such as the Leiden/Argentine/Bonn (LAB) survey, provide high-resolution, velocity-resolved data that allow direct comparison with theoretical models of the Galactic structure. 21 cm line astrophysics is quite rich and broad. In Chapter 3, we saw the instrumental and observational aspects of the HI line. In this chapter, we focus on HI observations to derive the Galactic rotation curve, compare features with LAB survey data, and infer the rotation curve and the location of the spiral arms. We look at the detection history and glimpse the important role HI plays in Galactic astrophysics. Mathematical details are presented for students interested in an introduction to galactic dynamics. It provides a pedagogical foundation for interpreting the observed data and constructing physical models of the Milky Way. Jupyter notebook tutorials are available for this chapter along with supplementary material.

doi:10.1088/2514-3433/ae1223ch4

4.2 Overview

The name HI refers to neutral atomic hydrogen and follows a standard astronomical nomenclature for labeling ionization states of elements. H denotes elemental hydrogen, and I (Roman numeral one) denotes the first ionization state, which, by convention, means neutral (not ionized). By extension, HII would mean singly ionized hydrogen, and so on. In this section, a brief overview of the HI emission physics, astronomy, and historical developments are presented.

4.2.1 HI Emission Physics

In the ground state of neutral hydrogen, the electron and proton spins can be either parallel (triplet state) or antiparallel (singlet state). The triplet state has slightly higher energy than the singlet state, with an energy difference of approximately 5.9×10^{-6} electron volts (Figure 4.1). The transition from the higher to the lower energy state results in the emission of a photon at 1420.40575 MHz. This transition is highly forbidden, with a spontaneous emission rate of about 2.9×10^{-15} s^{-1}, which corresponds to a mean lifetime of around 11 million years. Although the probability of the transition is small and hence the inferred lifetime for the transition is very long, an important aspect which plays a role is collisional excitation and de-excitation. In addition to the abundance of hydrogen, collisional de-excitation makes the detection of the line possible.

An important parameter that determines the rate of collision is the kinetic temperature T_K. The kinetic temperature is the physical temperature of the hydrogen gas. It measures the average thermal motion of the atoms and is defined by the Maxwell–Boltzmann distribution.

$$\frac{1}{2}m_H\langle v^2\rangle = \frac{3}{2}k_B T_K \tag{4.1}$$

where k_B is the Boltzmann constant, v is the velocity of the atoms and m_H is mass of the hydrogen atom. T_K contributes to the line broadening along with turbulence and the overlap of multiple Gaussians along the line of sight. For a gas of hydrogen atom at kinetic temperature T_K, the 1D velocity distribution along the line of sight is

$$f(v) = \sqrt{\frac{m_H}{2\pi k_B T_K}}\exp\left(-\frac{m_H v^2}{2k_B T_K}\right) \tag{4.2}$$

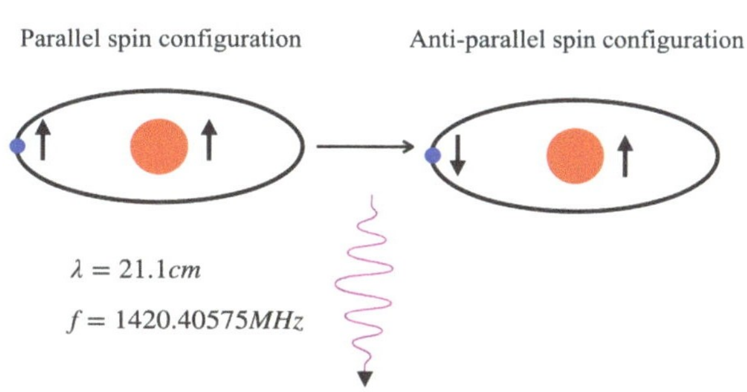

Figure 4.1. Spin flip transition of electron-proton pair in the hydrogen atom emits a 21 cm wavelength photon.

This is a Gaussian distribution at zero velocity with standard deviation,

$$\sigma_{\rm V} = \sqrt{\frac{k_{\rm B} T_{\rm K}}{m_{\rm H}}} \tag{4.3}$$

For the cold neutral medium (CNM) $T_{\rm K} \sim 50$–100 K or the warm neutral medium (WNM) $T_{\rm K} \sim 6000$–10000 K. Between the CNM and WNM, σ_v varies from 0.64 km s^{-1} to 9 km s^{-1}. In the observations we present, we see larger line widths, which is due to turbulence effects in the hydrogen clouds and overlap of multiple Gaussians along the line of sight.

While the kinetic temperature contributes to the line width, the spin temperature governs the brightness or line strength. The spin temperature ($T_{\rm s}$) determines the population ratio between the hyperfine levels, which in turn governs the intensity of the emission or absorption of the hydrogen line.

$$\frac{n_1}{n_0} = \frac{g_1}{g_2} \exp\left(-\frac{h\nu}{k_{\rm K} T_{\rm s}}\right) \tag{4.4}$$

where $g_1 = 3$ and $g_0 = 1$ are degeneracies and h is Planck's constant. If $T_{\rm back}$ is the background temperature, then $T_{\rm s}$ relative to $T_{\rm back}$ determines whether the line appears in emission or absorption. If $T_{\rm s} > T_{\rm back}$, the line appears in emission. If $T_{\rm s} < T_{\rm back}$, it appears in absorption. And if $T_{\rm s} = T_{\rm back}$, there is no contrast between the line and the background and it is not visible. $T_{\rm s}$ is influenced by collisions through which it couples to $T_{\rm K}$ and radiative transitions through which it couples to $T_{\rm back}$.

The observed brightness temperature, which is the temperature of the line relative to the background temperature, in the optically thin limit, i.e., the optical depth $\tau < <1$ is,

$$T_{\rm B} = (1 - e^{-\tau})(T_{\rm s} - T_{\rm back}) \approx \tau(T_{\rm s} - T_{\rm back}) \tag{4.5}$$

The temperatures mentioned in Figures 4.3 and 4.6 are brightness temperatures.

4.2.2 History of HI Line Detection

During World War II, Jan Oort, a prominent Dutch astronomer, recognized the limitations imposed by dust extinction on optical observations. He speculated about the possibility of detecting spectral lines at radio wavelengths, which would not be obstructed by the dust.

In Leiden in 1944 April, Oort posed the question of whether any such detectable radio spectral lines might exist. Hendrick van de Hulst, then a student, took up the challenge. Based on the understanding of the hyperfine structure of the hydrogen atom's ground state (the energy difference between states where the electron and proton spins are parallel versus antiparallel), van de Hulst calculated that a transition between these levels should occur. He predicted this transition would emit radiation with a wavelength of approximately 21 cm (frequency ≈ 1420.405 MHz; Van De Hulst 1982; Oort 1940). Although the transition probability (Einstein A-coefficient) was known to be extremely low (implying a lifetime of

$\sim$11 million years for the upper state), Hulst reasoned that the sheer abundance of hydrogen in interstellar space could make the cumulative signal detectable with the developing radio technology.

World War II spurred the development of microwave radio technology for radar research. After the war, this technology provided the necessary tools for radio astronomy. The field of radio astronomy saw a boom after Jansky discovered radio signals from celestial objects. Harold "Doc" Ewen and Edward M. Purcell achieved the first successful detection of the 21 cm hydrogen line at Harvard University, USA. Using a pyramidal horn antenna pointing out of a window at the Lyman Laboratory, they definitively detected the 21 cm emission from the Milky Way on 1951 March 25 (Ewen & Purcell 1951). Ewen used a frequency switching technique, which switched between the hydrogen line frequency (1420.405 MHz) and a nearby frequency for background subtraction.

Before publishing their results, Ewen and Purcell shared their techniques and findings with the Netherlands group. C. Alexander Muller and Jan Oort in the Netherlands used Ewen's frequency switching technique and detected the line on 1951 May 11 (Muller & Oort 1951), independently confirmed the detection within a few weeks. In Australia, W. N. Christiansen and J. V. Hindman at the Commonwealth Scientific and Industrial Research Organisation (CSIRO) facility in Potts Hill, Sydney, also detected the line shortly after the Dutch confirmation, reporting their results in mid-1951 (Christiansen & Hindman 1952). Ewen and Purcell did not rush to publish their results before sharing them with others. This serves as a notable example of scientific collaboration and mutual trust, prioritizing shared progress over competition.

4.2.3 HI Astronomy

The immediate and most profound impact of the 21 cm line detection was its application to mapping the structure of the Milky Way. By measuring the Doppler shifts (due to the rotation of the Galaxy) of the 21 cm emission from different directions, astronomers could determine the line-of-sight velocities and estimate the distances to hydrogen clouds. Combined efforts by the Dutch group in Leiden (led by Oort, Muller, and Gart Westerhout) and the Australian group in Sydney (led by Frank Kerr), produced the first large-scale maps of the neutral hydrogen distribution in the Milky Way during the 1950s (Oort et al. 1958). These maps provided the first observational evidence for the spiral structure of our Galaxy, revealing segments of spiral arms that are not seen in optical wavelengths due to dust. They also allowed for detailed studies of Galactic rotation. The following sections will use some of these methods to determine the rotation curve and Galactic structure from the HI line observations.

An important parameter related to the observable brightness temperature $T_{\rm B}$ is the column density $N_{\rm HI}$. It is defined as the number of number of particles, number of HI atoms in this case, per unit area along the line of sight. For optically thin case, $N_{\rm HI}$ is directly related to $T_{\rm B}$ as

$$N_{\rm HI} = 1.823 \times 10^{18} \int T_{\rm B}(v)dv \ [\rm cm^{-2}] \tag{4.6}$$

The integral is over the velocity profile of the 21 cm line [km s^{-1}]. For optically thick case, $N_{\rm HI}$ is given by,

$$N_{\rm HI} = 1.823 \times 10^{18} \int T_{\rm s}(v) \cdot \tau(v)dv \ [\rm cm^{-2}] \tag{4.7}$$

where $\tau(v)$ is the optical depth as a function of velocity. This is usually the case in the Galactic plane where $T_{\rm B} \geqslant 80$ K where we encounter cold dense HI clouds. Absorption studies can give an independent measure of $\tau(v)$. HI line can be observed in absorption if there is a strong background source like the Galactic center or a bright galaxy in the background. Since these background sources are usually compact and much smaller than the beam of the BHARAT antenna (discussed in Part I), it is difficult to see the HI line in absorption with the current setup. Typical values of $N_{\rm HI}$ are of the order of 10^{21} $\rm cm^{-2}$ (Dickey & Lockman 1990). The absorption studies of HI line can give important insights into galactic dynamics as absorption lines typically have lower width, hence higher velocity resolution. Absorption against bright background sources reveals cold, small-scale structures invisible in emission and allows direct estimation of gas spin temperatures (Dickey et al. 2003). Additional information comes from absorption-line studies and phenomena like high-velocity clouds (HVCs). HVCs and intermediate-velocity clouds (IVCs) are probes of galactic accretion or outflow, potentially indicating infall from the intergalactic medium or material recycled via supernova-driven fountains (Wakker & van Woerden 1997).

Beyond mapping spiral arms and defining the Milky Way's rotation curve, 21 cm hydrogen line observations provide insight into the interstellar medium (ISM). These measurements reveal the mass and vertical distribution of neutral hydrogen (HI), including features like the galactic warp and HI scale height. These estimates help compare gas to stellar and molecular components and inform models of galaxy evolution and star formation efficiency (Kalberla & Kerp 2009; Dickey & Lockman 1990). The velocity resolution of the 21 cm line also enables detection of non-circular motions caused by spiral arm streaming, bar dynamics, and gas inflows or outflows near star-forming regions (Kalberla & Kerp 2009). Deviations from circular rotation, warps, and velocity dispersion derived from line widths expose the turbulence, temperature, and dynamic processes affecting gas clouds (Dickey & Lockman 1990). By decomposing emission profiles, it is possible to distinguish between the Cold and Warm Neutral Medium (CNM/WNM; Heiles & Troland 2003).

Measurements of Zeeman splitting in the 21 cm line enable direct assessments of magnetic field strength within HI clouds, one of the few observational tools for probing Galactic magnetism at these scales (Heiles & Troland 2003, 2004). These applications make the 21 cm line an indispensable tool for studying the structure, dynamics, thermodynamics, and magnetism of the Milky Way's neutral ISM.

HI astronomy is foundational in understanding galaxies beyond the Milky Way. HI gas typically extends beyond the optical disc of the galaxies tracing galactic structure, morphology and rotation curves. It allows us to study interactions and mergers via HI bridges and streams (Walter et al. 2008). In recent years, the 21 cm line has become an important tracer for the early stages of the universe, just after the CMB formation. Due to the high redshift, the 21 cm line is shifted down to low frequencies. Several experiments are trying to measure the global 21 cm cosmological signal (de Lera Acedo et al., 2022; Singh et al., 2018; Bowman et al., 2018). The detection poses a challenge, since the signal is extremely faint compared to the foreground contamination. If detected, this will open up a new window to cosmology and advance the field of observational astronomy to a new paradigm.

4.3 Comparison with LAB Survey

Leiden/Argentine/Bonn (LAB) Galactic HI survey (Kalberla et al. 2005) is a high-sensitivity all-sky HI survey. It is an entire sky survey with 0.5° resolution. The survey is a combination of observations from three observatories that bear its name. We can use the survey to get the expected HI brightness from our antenna. The survey data cube can be downloaded.[1] Figure 4.2 shows one of the velocity slices of the data cube. We see that most of the HI is confined to the Galactic disc, i.e., 0 galactic latitude. The HI brightness changes with each velocity frame. Some features extend farther from the galactic disc, but there is little HI in the direction of the Galactic poles. Even then, this HI is detectable and is discussed in the calibration section of Chapter 3.

The antennas that we use have a much broader beam compared to the resolution of the survey. So, we can treat the LAB images as a close approximation of the true HI brightness in the sky. We are measuring the brightness temperature of the sky T_{sky} through the antenna beam **B**, so the sky brightness is convolved with the antenna

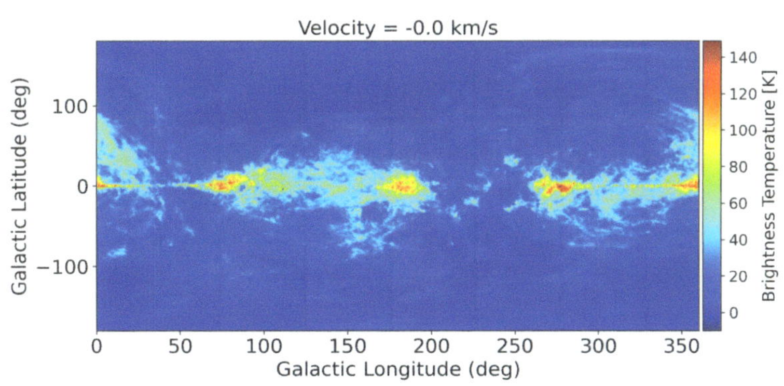

Figure 4.2. The figure shows a frame from LAB survey data cube. This is a slice of the Galaxy at zero velocity of the 21 cm line. We see that most of the bright features are near the Galactic plane, indicating that most of the HI in the Galaxy is confined to the Galactic disc. There is very little, but detectable, amount of neutral hydrogen outside the Galactic plane. We encounter this out-of-the-plane hydrogen in the cold sky observations discussed in Chapter 3.

[1] https://cdsarc.cds.unistra.fr/ftp/VIII/76/

beam at the output. Let us assume that the antenna beam has very low side lobes and we can approximate the antenna beam to be of a Gaussian shape with unity amplitude.

$$B(\theta) = \exp\left(-\frac{\theta^2}{2\sigma^2}\right) \tag{4.8}$$

Here θ is the angle coordinate of the beam and σ is the angular standard deviation or the spread of the antenna beam. We can now get the observed brightness from the sky brightness as follows.

$$T_{\text{obs}} = \int_{-\Delta\theta}^{\Delta\theta} T_{\text{sky}}(\theta')B(\theta - \theta')\, d\theta' \tag{4.9}$$

The limits of integration $\Delta\theta$ can be assumed to be 3σ in the present case to cover 99% of the Gaussian. We have assumed that the beam is symmetric about its axis. This may not always be the case. For precise measurements, the full description of beam B should be used, including the side lobes and the full solid angle limits for integration. Some features in the comparison may not be a perfect match due to incomplete beam description or antenna pointing errors. It should also be noted that the survey has a finite resolution and the pixels at the edge of a beam with a sharp cut-off will not be weighted correctly.

For the BHARAT (Mhaske et al. 2022) antenna, the Gaussian approximation is sufficient for the current experiment due to its low side lobes and circular beam. For the BHARAT antenna, the FWHM is 16.5° (see Chapter 3), which translates to σ of 7°. The effect of a larger beam is smoothing or weighted averaging of the true sky brightness.

Since our observations are in the Galactic plane (with 5° longitude spacing), we can create a velocity–longitude plot from the LAB survey as shown in Figure 4.3. We can see some distinct features in the plots as separate curves. These curves are the spiral arms of the Galaxy. It is possible to reconstruct the spiral structure of the Galaxy by fitting the spiral arms to the velocity–longitude plot (Simonson 1976; Burton 1971). We see that the convolved LAB plot 2 in Figure 4.3, which is derived from the original LAB velocity–longitude plot 1 in Figure 4.3 and the observed plot 3 in Figure 4.3, is a good match.

The convolved LAB data can also be used to calibrate the system. We can get the expected brightness temperature from the survey and scale the observed plot accordingly to match the expected brightness and correct the velocity errors as well. This is another way to calibrate the plots apart from the method discussed in Chapter 3.

4.4 Galactic Dynamics

The field of Galactic dynamics seeks to understand the large-scale motions of stars, gas, and dark matter within galaxies, particularly the Milky Way, using the principles of gravitation and orbital mechanics. One of the earliest and most influential models was developed by Jan Oort, who applied stellar velocity

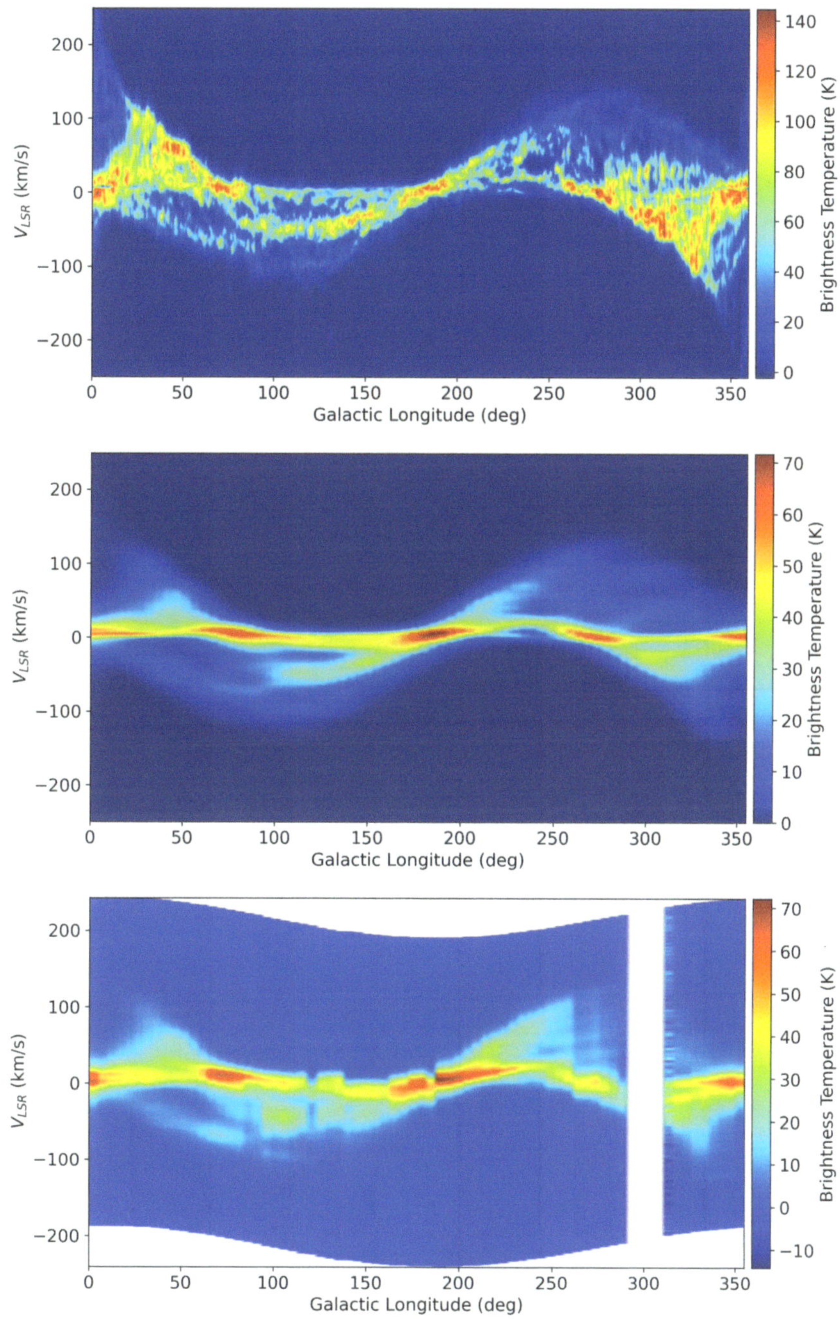

Figure 4.3. (Top to bottom) (1) Velocity–longitude plot from LAB survey, (2) velocity–longitude plot from Gaussian weighted LAB survey and (3) velocity–longitude plot from observations with the BHARAT setup. We see a good match between the Gaussian weighted LAB survey and the observation with the BHARAT setup. The minor deviations are attributed to pointing errors in the antenna. But the overall features are captured very well in the observations.

measurements to infer local Galactic rotation and introduced the Oort constants, which describe differential rotation near the Sun (Oort 1927). This foundational work established that the Galaxy exhibits organized rotation rather than random stellar motion.

Broadly the Galaxy is made up of the central bulge and the galactic disc. The central bulge is a dense spheroidal mass distribution up to a few kiloparsecs (kpc). The motion of stars in the bulge is mostly random compared to the near-circular orbits in the disc, which is rotation dominated extending to the edge of the visible edge of the Galaxy. It has be observationally determined that the rotation of the Galactic disc is differential, i.e., the angular velocity of the material in the disc is different for different radii. As we will see, we find the angular velocity is inversely proportional to the radius or velocity is constant, giving a flat rotation curve.

This section discusses the dynamics of the galaxy including the spiral arms, the components of the rotation curve and look into observational estimation of the rotation curve.

4.4.1 The Spiral Arms

The spiral arms are the luminous, often grand, patterns observed in the disks of a significant fraction of galaxies. They are sites of active star formation, characterized by higher densities of gas, dust, young blue stars, and HII regions than the inter-arm regions. Understanding how these structures form and persist over billions of years, despite the differential rotation of galactic disks, is a fundamental problem in galactic dynamics. The most immediate problem arises from galactic differential rotation. Stars and gas clouds in a galactic disk do not orbit at the same angular speed (Ω). Typically, $\Omega(R)$ decreases with increasing radius R from the galactic center. If spiral arms were simply material structures composed of the same stars and gas clouds over time, differential rotation would wind them up tightly as seen in Figure 4.4c.

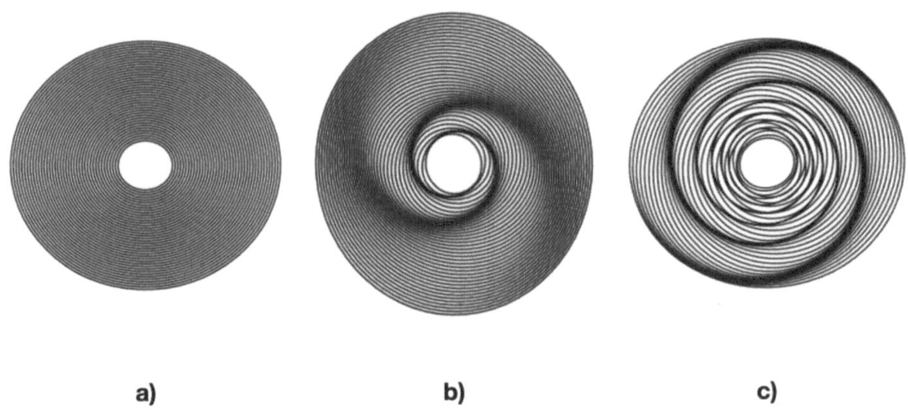

Figure 4.4. The figure shows that material arms can lead to spiral winding. (a) Elliptical orbits of objects in a galaxy, (b) the ellipse major axis edges form high-density regions due to differential rotation, (c) the spiral structure gets wound up and disappears in a time span much smaller than the age of a galaxy. This contradicts the observations, suggesting that the spirals are not material structures but could be density waves.

Consider two points on a material arm initially aligned radially, at radii R and $R + \Delta R$. Their angular velocities are $\Omega(R)$ and $\Omega(R + \Delta R)$. We can get

$$\Omega(R + \Delta R) \approx \Omega(R) + \frac{d\Omega}{dR}\Delta R \tag{4.10}$$

After a time t, the difference in azimuthal angle swept out is,

$$\Delta\theta = [\Omega(R) - \Omega(R + \Delta R)]t \approx -\left(\frac{d\Omega}{dR}\right)\Delta Rt \tag{4.11}$$

For a typical spiral galaxy like the Milky Way, $\frac{d\Omega}{dR}$ is negative (rotation curve is flat or falling in the outer parts). An arm segment spanning a few kiloparsecs would become tightly wound ($\Delta\theta \gg 2\pi$) in just a few hundred million years, much less than the age of the Galaxy (Binney & Tremaine 2008). Observations, however, show relatively open spiral patterns persisting for much longer timescales.

The winding problem requires spiral arms to be not purely material structures, but patterns propagating through the disk. Several theories attempt to explain their origin and persistence. The quasi-static density wave theory proposed by C. C. Lin and Frank Shu in the 1960s (Lin & Shu 1964; Shu 1970) is the most influential theory, particularly for explaining large-scale, grand-design spirals. According to this theory, spiral arms are quasi-static (long-lived) density waves or regions of enhanced density that move through the stellar and gaseous disk at a fixed angular speed, the pattern speed (Ω_p), which generally differs from the material rotation speed $\Omega(R)$. Stars and gas clouds orbit the Galaxy, pass through the slower-moving density wave, are compressed, and then move out again. The arm is the locus of this compression, analogous to a traffic jam on a highway where individual cars slow down. However, the jam persists or moves slowly (see simulation in Supplementary material).

This basic theory does not fully explain the origin and maintenance of the density wave. External gravitational perturbations or internal asymmetries can excite or amplify density waves. A close encounter or flyby with another galaxy can exert strong tidal forces, pulling out material and triggering the formation of spiral density waves, often resulting in prominent grand-design arms (Toomre & Toomre 1972). Simulations show this is a potent mechanism.

Many spiral galaxies possess a central stellar bar. These rotating, non-axisymmetric structures can effectively drive spiral density waves into the surrounding disk (Sellwood & Sparke 1988). The bar rotates with its pattern speed, and gravitational torques from the bar exert forces on the disk material. Resonances play a crucial role, particularly Lindblad Resonances (where the frequency of stellar epicyclic motion matches the relative frequency of the passing pattern) and the Corotation Resonance (where $\Omega(R) = \Omega_{p}$, bar). Bars effectively channel gas inwards and drive a spiral structure (Kormendy & Kennicutt 2004).

The spiral arms of a galaxy play a crucial role in galactic dynamics and galactic chemistry. Since they are regions of high density, they trigger star formation in the interstellar medium as it collapses while passing through these arms. We find a large population of young stars in spiral arms due to this effect. Stars eventually create heavy atoms, giving rise to the rich chemistry of the Galaxy.

4.4.2 Rotation Curve

The rotation curve gives the relation between the distance and velocity of the Galactic disc. It is an important measurement to understand the Galactic dynamics. In this section, we will look at the method of determining the rotation curve of the Milky Way with the HI observations. We will try to understand the rotation curve with galactic dynamics and compare the model with our observations.

4.4.3 Rotation Curve from Observations

It is evident from the observations of the HI line that most of the HI in the Galaxy is located in a disc; the disc rotates, which creates the Doppler shifts in the line and that there are spiral arms, which is indicated by the multiple bright peaks in the observations. It is possible to get the rotation curve of the Galaxy for radius less than the solar radius R_0 with HI observations. This method was implemented (Kwee et al. 1954; Oort et al. 1958). We will get the inner rotation curve in this section following similar arguments.

Consider Figure 4.5, which shows the orbit of the Sun around the Galactic center. The line of sight of an observation for Galactic longitude l crosses the arms of the Galaxy at several points. These appear as peaks in the line plot as seen in Figure 4.6. For circular orbits inside the solar radius, the line of sight crosses each arm at two points. These two peaks will overlap in the velocity–temperature plot of the HI line

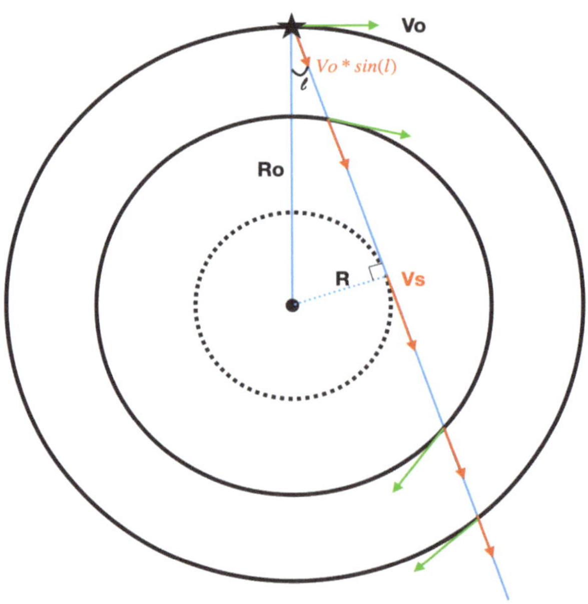

Figure 4.5. Schematic of the rotation of the Galaxy. We assume a clockwise rotation and circular orbits, which is a good enough approximation for the purpose here. In the method described here we can plot the rotation curve only for sources inside the Solar orbit around the Galactic center. R_o and V_o are the distance and velocity of the Sun with respect to the Galactic center, while l, R, and V_s are the Galactic longitude, distance and the velocity of the source with respect to the Galactic center.

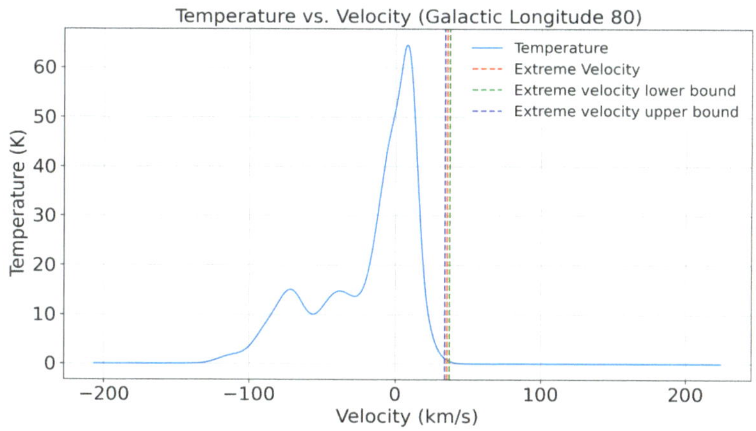

Figure 4.6. The model of the line created by fitting Gaussians to the line profile (see Chapter 3). We use the model here instead of the calibrated line profile because it is easier to choose the extreme point on the model compared to the noisy line profile. The extreme velocity (most red-shifted) at Galactic longitude 80 is chosen here by choosing a point 3σ (RMS noise) above zero. The upper and lower bounds are the $\pm 1\sigma$ values before and after the extreme velocity. These bounds are used to quantify the uncertainty in the measurement of the extreme velocity.

(Figure 4.6). This ambiguity prohibits us from determining the position and velocity of the points in the HI plot uniquely. This ambiguity breaks for only one point along the line of the site, i.e., the tangential point. This is a unique point in the HI line and in the schematic as well. The tangential point corresponds to the most red-shifted point in the plot for the first quadrant of the Galaxy (galactic longitudes 0° to 90°) and the most blue-shifted point in the fourth quadrant (galactic longitudes 270° to 360°). Since these are unique points in the plots, we can assign a unique distance and velocity to these points.

We can safely assume that the velocity for orbits closer to the Galactic center will be faster than the velocity of the Sun. A simple Keplerian rotation will satisfy this condition. We know the radius and velocity of the solar orbit through independent measurements, $R_0 = 8.5$ kpc and $V_0 = 235$ km s^{-1}. Therefore, the velocity that we measure V_{obs} in our observations is the relative velocity between the line-of-sight components of the Sun's velocity and the source velocity V_s.

$$V_{\text{obs}} = V_s - V_0 \sin(l) \tag{4.12}$$

From this equation, we get the following.

$$V_s = V_{\text{obs}} + V_0 \sin(l). \tag{4.13}$$

V_{obs} corresponds to the extreme velocity as seen in Figure 4.6.

We can also obtain R, the radius for this source, with simple trigonometry.

$$R = R_0 \sin(l). \tag{4.14}$$

The extreme velocity is chosen to be the point that is most red-shifted and is 3σ above the plot baseline. We put a lower and upper bound on the extreme velocity

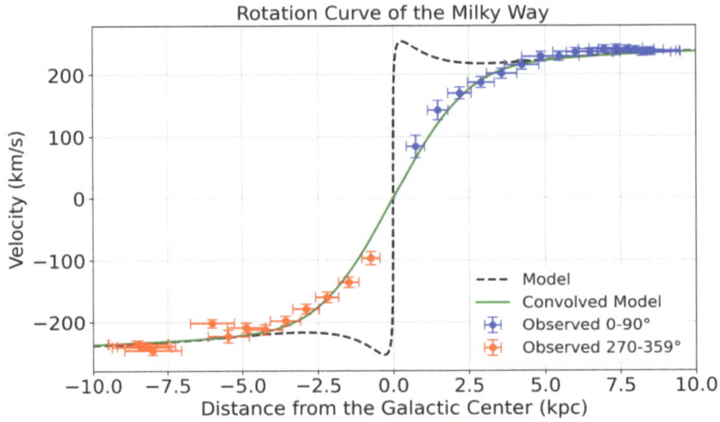

Figure 4.7. The figure shows the observed (with BHARAT setup) rotation curve of the Galaxy compared with the rotation curve measured using larger antennas (higher angular resolution). The rotation curve observed with larger telescopes is like a step at the Galactic center. If this step is smoothed out with a beam of the antenna, we find that it matches the observation from the BHARAT antenna indicating that the deviation from the high-resolution surveys near the center is due to the averaging caused by the large beam of the BHARAT antenna. The positive half of the rotation curve comes from the observation between 0° to 90° Galactic longitudes (the first quadrant) and the negative half of the curve is computed with the observations from between 270° to 359° Galactic longitudes.

with $\pm 1\sigma$ velocity points above and below the extreme velocity point. Finally, using Equations (4.13) and (4.14), we get the rotation curve of the inner Galaxy as shown in Figure 4.7.

The model will be discussed in the following section. We see that the measured velocities match the model very well near the solar radius. As we will see in the following section, the visible mass of the Galaxy is not sufficient to explain the high velocities that we measure as seen in Figure 4.10.

The large beam of the antenna can explain the deviation from the model near the Galactic center. The beam of the BHARAT antenna is about 16.5°. All sources within this beam will be convolved together. If we smooth out the model with a Gaussian of width of the beam, we get a curve that matches our observations (see Figure 4.7). The Galaxy's high-resolution rotation curve can be achieved using the same LAB survey method.

4.4.4 Components of Rotation Curve

The rotation curve of a galaxy is the measure of how fast the Galaxy is rotating as a function of distance. If we assume that the majority of the mass (M) in the Galaxy is in the center, then the gravitational force on a test mass (m) in a circular orbit around the Galactic center is

$$F = \frac{GMm}{r^2} \tag{4.15}$$

If we balance this force by the centripetal force for a stable orbit, then we get

$$\frac{GMm}{r^2} = \frac{mV^2}{r} \tag{4.16}$$

which gives the velocity

$$V = \sqrt{\frac{GM}{r}} \tag{4.17}$$

This is a very simple model of the Galaxy. From this *Keplerian* model, we see that the rotation velocity should fall as the inverse square root of the distance r. This is clearly not true from our observations and other observations of the rotation curve for our Galaxy (Clemens 1985) as well as other galaxies. To be more precise, let us assume that the mass is distributed in the Galaxy. Then the rotation curve of a galaxy depends on the mass distribution. We consider a three-component model of the Galaxy (Figure 4.8). These are discussed below.

(1) **The Bulge**

The bulge is the central, almost spherically shaped region of the Galaxy. Although there is evidence that the central bulge of the Milky Way is a bar-shaped structure, let us assume it to be spherical. We find an old star population in the bulge and very little cold gas. The bulge can be seen above and below the galactic plane in the optical wavelengths. When we take pictures of spherical distributions of stars, we get the 2D projection of a 3D distribution. The density of the projected profiles is called surface density. Since we measure luminosity, the 2D density is the luminosity surface density. A popular model for describing the projected spherical densities is the Sersic profile, which is an exponential in $R^{1/n}$ (R as radius). The value of n for bulges is 4. Hence, a special case of the Sersic profile, the de

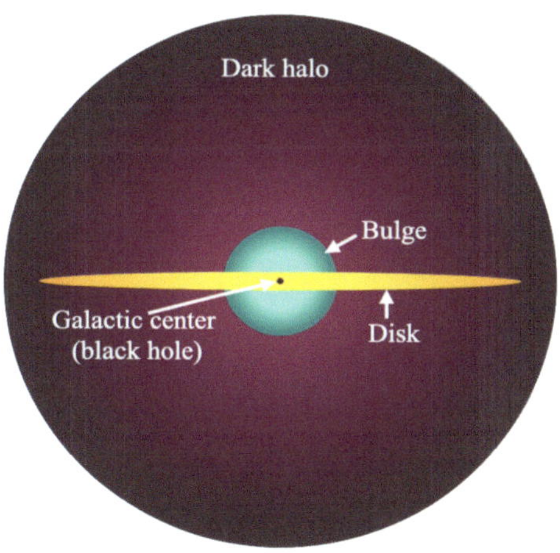

Figure 4.8. The mass components of the Milky Way Galaxy. We assume three major components of the mass distribution in the Galaxy, i.e., the central bulge, the Galactic disc and the dark halo. The black hole does not affect the large-scale gravitational potential of the Galaxy, which affects the overall motion of the objects in the galaxy. The bulge and the disc are the visible while the dark halo is the invisible hypothesized dark matter, which extends far beyond the visible disc of the Galaxy.

Vaucoulers profile, is also sometimes referred to as the $R^{1/4}$ law. If we know the mass-to-light ratio, we can convert the luminosity into mass. The de Vaucoulers profile is described below (Sofue 2015).

$$\Sigma_b(r) = \Sigma_{b0} \exp\left[-\kappa\left(\left(\frac{r}{R_b}\right)^{\frac{1}{4}} - 1\right)\right] \tag{4.18}$$

Here, $\kappa = 7.6695$ and Σ_{b0} is given by

$$\Sigma_{b0} = \frac{M_b}{\eta R_b^2} \tag{4.19}$$

where $\eta = 22.665$, R_b is the scale radius, and M_b is the scale mass. The problem now is to convert this surface mass density $\Sigma_b(r)$ into the 3D mass density of the bulge. To simplify things, we assume the distribution to be spherical. Consider the situation in Figure 4.9 where a spherical distribution

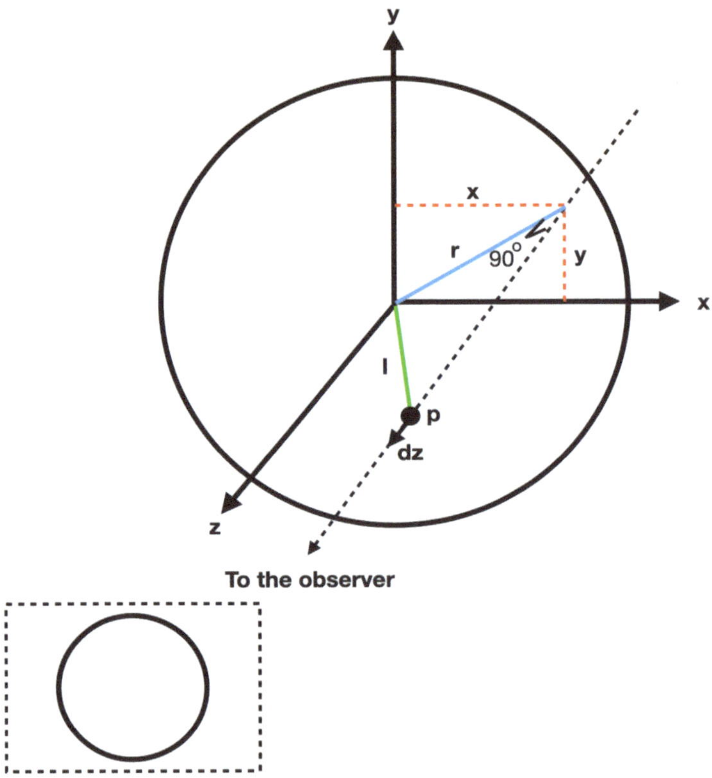

Figure 4.9. The figure shows the de-projection of the Milky Way bulge. The light from the stars that are spherically distributed is projected on to a photographic plate of the observing instrument. The 2D image has varying density profile which known as surface density. Each point on the photographic plate can be assumed to be formed by light coming from all the sources in the line perpendicular to the photographic plate at that point. The goal of de-projection is to reconstruct the 3D density profile of the sources from the 2D surface densities.

is projected onto a 2D plate. Since the projection is along the z-axis, each pixel on the plate is the integrated intensity along a line parallel to the z-axis. Let us consider one such line (dashed black line). Along this line, the x and y coordinates do not change. Consider a point p on this line with coordinates (x, y, z) at a distance l from the origin.

$$\begin{aligned} l &= \sqrt{x^2 + y^2 + z^2} \\ r &= \sqrt{x^2 + y^2} \\ \Longrightarrow l &= \sqrt{r^2 + z^2} \\ z &= \sqrt{l^2 - r^2} \\ \Longrightarrow \frac{dz}{dl} &= \frac{l}{\sqrt{l^2 - r^2}} \end{aligned} \tag{4.20}$$

We can get $\Sigma_b(r)$ (2D density) by integrating $\rho(l)$ along the line.

$$\Sigma_b(r) = \int_{-\infty}^{\infty} \rho(l)\, dz \tag{4.21}$$

The minimum value of l is r as l moves along the line, as seen in Figure 4.9. Also, the integral is symmetric about the x–y plane. Substituting dz from Equation (4.20), we get

$$\Sigma_b(r) = 2\int_{r}^{\infty} \rho(l)\frac{dz}{dl}\, dl \tag{4.22}$$

$$\Sigma_b(r) = 2\int_{r}^{\infty} \rho(l)\frac{l}{\sqrt{l^2 - r^2}}\, dl \tag{4.23}$$

Equation (4.24) is an Abel integral. It can be inverted as follows in Appendix B.5 of Binney & Tremaine (2008):

$$\rho(l) = -\frac{1}{\pi}\int_{l}^{\infty} \frac{d\Sigma_b(r)}{dr}\frac{dr}{\sqrt{r^2 - l^2}} \tag{4.24}$$

where

$$\frac{d\Sigma_b(r)}{dr} = -\frac{\kappa\Sigma_{b0}}{4R_b}\left(\frac{r}{R_b}\right)^{-\frac{3}{4}} \exp\left[-\kappa\left(\left(\frac{r}{R_b}\right)^{\frac{1}{4}} - 1\right)\right] \tag{4.25}$$

using Equation (4.18).

Now that we have the spherical density $\rho(l)$ profile, we can find the mass enclosed inside radius R as follows. Imagine a thin shell of area $4\pi r^2$ and thickness dr. We can then integrate to get $M_b(R)$ as

$$M_b(R) = 4\pi\int_{0}^{R} r^2\rho(r)\, dr. \tag{4.26}$$

Once the mass enclosed inside radius R is known, we can find velocity of a star at radius R using same argument as Equations (4.16) and (4.17)

$$V_b(R) = \sqrt{\frac{GM_b(R)}{R}} \tag{4.27}$$

Equations (4.24), (4.26), and (4.27) can be solved numerically to get the bulge contribution to the Galactic rotation curve in Figure 4.10 (see tutorials for the code). We observe a steep rise in the rotational velocity in the inner few kiloparsecs of the bulge. This is due to the concentrated mass distribution near the Galactic center, which is tightly bound gravitationally and results in a motion akin to a solid body, i.e., velocity increases linearly with radius.

(2) **The Disc**

The Milky Way is a spiral galaxy with a disc made of spiral arms. The disc dynamics and structure will be discussed in detail in the Galactic Structure section. The mass density of the disc is exponential.

$$\Sigma_d(R) = \Sigma_{d0} \exp\left(\frac{-R}{R_d}\right) \tag{4.28}$$

where Σ_{d0} is the central value of the mass density and R_d is the scale radius. The derivation of the rotation curve of the exponential disc can be found in Binney & Tremaine (2008) and Freeman (1970) and is given by the following expression. (See Appendix: A for derivation.)

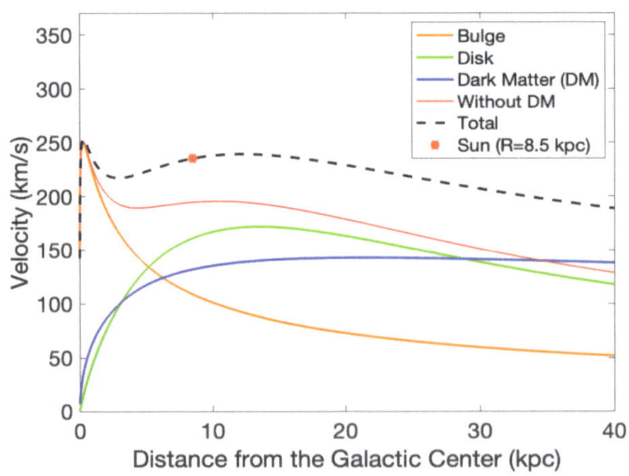

Figure 4.10. The figure shows the components of the rotation curve of the Milky Way. It is evident that the total observed rotation curve (dashed line) cannot be recovered by considering just the central bulge and the disk of the Galaxy. The additional gravitational potential comes from the invisible gravitational mass popularly known as dark matter.

$$V_d(R) = \sqrt{4\pi G\Sigma_{d0}R_d y^2[I_0(y)K_0(y) - I_1(y)K_1(y)]} \tag{4.29}$$

Here $y = \frac{R}{2R_d}$ and I and K are modified Bessel functions.

The motion of matter in the disc is dominated by rotation and is almost circular.

(3) **The Halo**

The evidence of extra mass in the Galaxy is clearly seen in Figure 4.10. Visible baryonic matter is not sufficient to explain the large velocity of the Galactic disc, especially at the outer edge of the Galaxy. This effect is observed in other galaxies as well (Sofue & Rubin 2001). The dark halo, also known as dark matter, is seen in galaxy clusters. Fritz Zwicky first observed in the Coma cluster (Zwicky 1937) that the baryonic mass was insufficient to hold the galaxies. Later, the presence of dark halo was also observed in galaxy rotation curves by Jan Oort with HI observation (Oort 1932) and Vera Rubin and Kent Ford (Rubin et al. 1980) by observing stellar motions in the Andromeda galaxy.

Several models exist for the density distribution of dark matter in the Galaxy. The model used here was proposed in the 1990s by Julio Navarro, Carlos Frenk, and Simon White, known as the NFW model, and is based on the Cold Dark Matter (CDM) model for dark matter (Navarro et al. 1996). According to the CDM model, the dark matter is made of non-relativistic particles that interact weakly with electromagnetic radiation. The NFW model is developed from numerical simulations of structure formation in the Universe dominated by cold dark matter. The model is described as follows.

$$\rho_{\mathrm{NFW}}(r) = \frac{\rho_0}{\left(\frac{r}{R_h}\right)^{\alpha}\left(1 + \frac{r}{R_h}\right)^{\beta-\alpha}} \tag{4.30}$$

ρ_0 and R_h is the scale density and scale radius. The value of β is 3, and α is between 1 and 1.5. We use $\alpha = 1$. The mass enclosed in radius R can be found as follows. The density distribution is spherical as the density depends only on r. We imagine a shell of area $4\pi r^2$ with width dr. Therefore, the mass in volume $4\pi r^2 dr$ is

$$\begin{aligned}
dM_{\mathrm{dm}} &= \rho_{\mathrm{NFW}}(r)4\pi r^2 dr \\
M_{\mathrm{dm}}(R) &= \int_0^R \rho_{\mathrm{NFW}}(r)4\pi r^2\, dr \\
M_{\mathrm{dm}}(R) &= 4\pi\rho_0 \int_0^R \frac{r^2}{\frac{r}{R_h}\left(1 + \frac{r}{R_h}\right)^2}\, dr.
\end{aligned} \tag{4.31}$$

Substituting $x = \frac{r}{R_h}$,

$$M_{\rm dm}(R) = 4\pi\rho_0 R_h^3 \int_0^x \frac{x}{(1+x)^2}\,dx \tag{4.32}$$

Consider only the integral part, if $u = 1 + x$,

$$\int_1^{1+x} \frac{u-1}{u^2}\,du = \int_1^{1+x}\left[\frac{1}{u} - \frac{1}{u^2}\,du = \left[\ln(u) + \frac{1}{u}\right]_1^{1+x} = \ln(1+x) + \frac{1}{1+x} - 1\right]_0^{R/R_h} \tag{4.33}$$

Substituting the limits, we get

$$M_{\rm dm}(R) = 4\pi\rho_0 R_h^3\left[\ln\left(1 + \frac{R}{R_h}\right) - \frac{R}{R+R_h}\right]. \tag{4.34}$$

Again, using the argument from Equation (4.16) and (4.17), we get a velocity contribution of dark matter halo $V_{\rm dm}$ as,

$$V_{\rm dm}(R) = \sqrt{\frac{GM_{\rm dm}(R)}{R}} \tag{4.35}$$

4.4.4.1 Rotation Curve

The three components can now be combined to get the full rotation curve of the Galaxy as follows.

$$V = \sqrt{V_b^2 + V_d^2 + V_{\rm dm}^2} \tag{4.36}$$

The parameters for the model presented here are taken from Sofue (2015), with slight variation (within error bars), to fit the rotation curve observed in this experiment. These are given in Table 4.1.

Table 4.1.

Component	Parameter	Value
Bulge	M_b	0.25×10^{11} $M_{\rm sun}$
	R_b	0.87 kpc
Disc	M_d	1.12×10^{11} $M_{\rm sun}$
	R_d	6.3 kpc
Halo	ρ_0	18.2×10^6 $M_{\rm sun}$
	R_h	9.8 kpc

The Galactic center is a super-massive black hole of mass 4×10^6 M_{sun}. The gravitational influence of the black hole is a small fraction compared to the bulge mass.

Currently, there is more than one model for each component, especially for the dark halo (Sofue 2020). There are only two parameters in the three components: length and mass. These values are obtained by fitting the observed rotation curve. Since there are large error bars on the observations of the rotation curves, this leads to degeneracy in the three components. The error bars for the outer edge of the Galaxy are large, making it difficult to separate out the components, especially the disc and halo components, as there are several values of the parameters, as well as several models that can fit the rotation curve. More precise measurements are required to narrow down the models of the components.

Some current problems in the rotation curve are to find how far the dark halo extends and what kind of density profile best predicts the rotation curve. There is also the possibility of modified gravity (Modified Newtonian Dynamics or MOND for short) to explain the rotation curves and rule out dark matter altogether.

4.5 Galactic Structure

The use of the 21 cm hyperfine transition of neutral hydrogen (HI) as a tool for Galactic astronomy was first demonstrated by Jan Oort and collaborators in the early 1950s, shortly after the initial detection of the line by Ewen and Purcell. Oort recognized that HI is observable from the far reaches of the Galaxy enabling us to study large-scale kinematics and morphology. Oort applied the new 21 cm data to extend his earlier models of Galactic rotation and introduced refined techniques to derive the Galactic rotation curve and structure from HI emission (Oort 1940; Oort et al. 1958).

4.5.1 Galactic Structure from Observations

Since we know that the rotation curve is flat, we can use the following arguments to get the distribution of HI in the Galactic plane. We need to find angles α and radial distance R in Figure 4.11 to plot the distribution of HI in polar coordinates. We can use the peak velocities V_{peak} from the Gaussian model of the line to trace the location of the brightest sources of HI.

In Figure 4.11, the angles l (Galactic longitude), α, and θ form a triangle. We can use the sine identity for the three angles to get,

$$\frac{\sin(\theta)}{R_o} = \frac{\sin(l)}{R} \implies \sin(\theta) = \left(\frac{R_o \sin(l)}{R}\right) \tag{4.37}$$

The velocity of an observed Gaussian peak is the relative velocity between the source velocity and the observer's (Sun) velocity along the line-of-sight.

$$V_{\text{peak}} = V_o \sin(\theta) - V_o \sin(l) \implies \sin(\theta) = \frac{V_{\text{peak}} + V_o \sin(l)}{V_o} \tag{4.38}$$

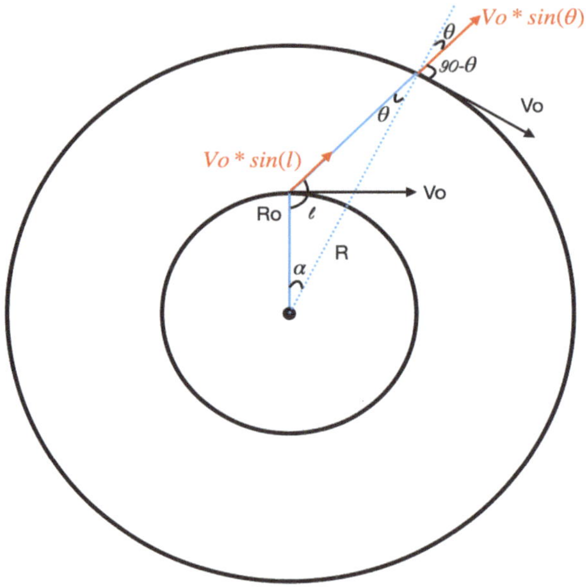

Figure 4.11. Schematic for deriving the structure of the Milky Way. The inner and the outer circles are the orbits of the Sun and a typical source, respectively. We assume a clockwise rotation.

Equating the $\sin\theta$ in Equation (4.37) and (4.38), we get R as,

$$R = \frac{V_o R_o \sin(l)}{V_{\text{peak}} + V_o \sin(l)} \tag{4.39}$$

We can substitute R back in Equation (4.37) to get θ and subsequently the polar angle α.

$$\theta = \sin^{-1}\left(\frac{R_o \sin(l)}{R}\right) \tag{4.40}$$

$$\alpha = \pi/2 - \theta - l \tag{4.41}$$

Once we have R and α, we can plot the location of the Gaussian peaks in Figure 4.12. We can see that the HI peaks follow a spiral structure. The evidence of the existence of the spiral arms in the Galactic disc is verified through these observations. For $R < R_{\text{sun}}$, θ can have two values, i.e., there is ambiguity in the location of the peak. In Figure 4.12 only one point is plotted.

We see distinct spiral arms that roughly trace the stellar arms of the Galaxy, although, the stellar and gas components have different dynamics and result in different distribution in the disc (Van De Hulst 1982). The HI distribution has more features compared to the stellar distribution. This demonstrates the usefulness of HI observations for studying far reaches of the Galaxy.

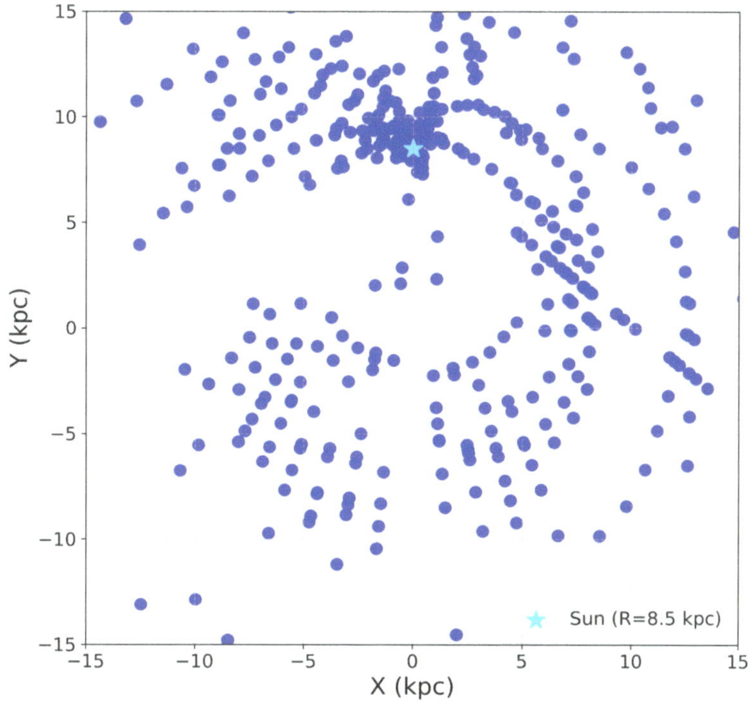

Figure 4.12. The structure of the Milky Way derived from the observed HI line profiles. We can clearly see the distinct spiral arms. The peaks of the Gaussians trace the spiral structures in the Galaxy. The straight line feature seen in the right central portion of the figure can be attributed to overestimation of the number of Gaussian peaks in a line profile in presence of noise and system nonlinearities. Some portion of the points are missing on the left central part of the figure. These are the directions that are not visible from the telescope location.

4.6 Conclusion

We have seen the rich astrophysics that can be derived from HI detection. This chapter has examined the role of neutral hydrogen (HI) in the discovery of the structure and dynamics of the Milky Way, beginning with its historic detection in 1951 by Ewen and Purcell. Using 21 cm line emission, we explored how HI traces the Galactic disk and halo, enabling the construction of rotation curves that reveal the presence of dark matter. We derived the three-component model of the Galaxy and analyzed the origin of spiral arms. Pedagogically, this chapter connects theoretical derivations with real survey data, offering students practical tools to interpret HI spectra, model Galactic kinematics, and understand the large-scale morphology of the Galaxy. It serves as both a scientific synthesis and a foundation for deeper exploration in Galactic astronomy. It is notable that the HI astronomy is a thriving field with several cutting-edge problems, some of which were hinted at in this chapter.

4.7 Appendix A: Derivation of Circular Velocity for an Exponential Disk

The circular velocity $V_d(R)$ in the plane of an infinitely thin disk with an exponential surface density $\Sigma(R) = \Sigma_{d0}e^{-R/R_d}$ is derived here in full for reference (Binney & Tremaine 2008; Freeman 1970). The gravitational potential $\Phi(R, z)$ is determined using Poisson's equation, and the Hankel transform is applied to handle the axisymmetric geometry.

The gravitational potential $\Phi(R, z)$ satisfies Poisson's equation.

$$\nabla^2\Phi(R, z) = 4\pi G\rho(R, z) \tag{4.42}$$

For an infinitely thin disk lying in the plane $z = 0$, the mass density is

$$\rho(R, z) = \Sigma(R)\delta(z) \tag{4.43}$$

where $\Sigma(R)$ is the surface density, and $\delta(z)$ is the Dirac delta function. Thus, Poisson's equation becomes

$$\nabla^2\Phi(R, z) = 4\pi G\Sigma(R)\delta(z) \tag{4.44}$$

Since the disk is axisymmetric, we express the Laplacian in cylindrical coordinates (R, z):

$$\nabla^2\Phi = \frac{1}{R}\frac{\partial}{\partial R}\left(R\frac{\partial\Phi}{\partial R}\right) + \frac{\partial^2\Phi}{\partial z^2} \tag{4.45}$$

Substituting into Poisson's equation:

$$\frac{1}{R}\frac{\partial}{\partial R}\left(R\frac{\partial\Phi}{\partial R}\right) + \frac{\partial^2\Phi}{\partial z^2} = 4\pi G\Sigma(R)\delta(z) \tag{4.46}$$

To solve this, we use the zeroth-order Hankel transform for axisymmetric functions in cylindrical coordinates. The transform is defined as:

- **Forward transform**: $\tilde{f}(k) = \int_0^\infty Rf(R)J_0(kR)\, dR$
- **Inverse transform**: $f(R) = \int_0^\infty k\tilde{f}(k)J_0(kR)\, dk$

where J_0 is the Bessel function of the first kind, order zero. We apply this to $\Phi(R, z)$ and $\Sigma(R)$.

$$\begin{aligned}\Phi(R, z) &= \int_0^\infty k\tilde{\Phi}(k, z)J_0(kR)\, dk \\ \Sigma(R) &= \int_0^\infty k\tilde{\Sigma}(k)J_0(kR)\, dk\end{aligned} \tag{4.47}$$

to transform Poisson's equation, the Hankel transform of the radial term $\frac{1}{R}\frac{\partial}{\partial R}(R\frac{\partial\Phi}{\partial R})$ becomes $-k^2\tilde{\Phi}(k, z)$ (Binney & Tremaine 2008), and the z-derivative term $\frac{\partial^2\Phi}{\partial z^2}$ transforms to $\frac{\partial^2\tilde{\Phi}}{\partial z^2}$. The right-hand side transforms as

$$4\pi G\Sigma(R)\delta(z) \rightarrow 4\pi G\tilde{\Sigma}(k)\delta(z) \tag{4.48}$$

Thus, in k-space, Poisson's equation is

$$\frac{\partial^2\tilde{\Phi}(k, z)}{\partial z^2} - k^2\tilde{\Phi}(k, z) = 4\pi G\tilde{\Sigma}(k)\delta(z) \tag{4.49}$$

This is an ordinary differential equation in z for each k. For $z \neq 0$, where the right-hand side is zero:

$$\frac{\partial^2\tilde{\Phi}}{\partial z^2} - k^2\tilde{\Phi} = 0 \tag{4.50}$$

The general solution is

$$\tilde{\Phi}(k, z) = A(k)e^{kz} + B(k)e^{-kz} \tag{4.51}$$

Since the potential must remain finite as $|z| \rightarrow \infty$:

- For $z > 0$, we need e^{-kz} (since $k > 0$).
- For $z < 0$, we need e^{kz}.

Given the symmetry of the disk, $\tilde{\Phi}(k, z)$ is even in z, so we write:

$$\tilde{\Phi}(k, z) = C(k)e^{-k|z|} \tag{4.52}$$

To determine $C(k)$, integrate the ODE across $z = 0$ from $-\varepsilon$ to ε and take the limit $\varepsilon \rightarrow 0$:

$$\int_{-\varepsilon}^{\varepsilon}\left(\frac{\partial^2\tilde{\Phi}}{\partial z^2} - k^2\tilde{\Phi}\right)dz = \int_{-\varepsilon}^{\varepsilon} 4\pi G\tilde{\Sigma}(k)\delta(z)\, dz \tag{4.53}$$

- Right-hand side: $4\pi G\tilde{\Sigma}(k)\int_{-\varepsilon}^{\varepsilon} \delta(z)\, dz = 4\pi G\tilde{\Sigma}(k)$
- Left-hand side: The term $-k^2\int_{-\varepsilon}^{\varepsilon} \tilde{\Phi}\, dz \rightarrow 0$ as $\varepsilon \rightarrow 0$ (since $\tilde{\Phi}$ is continuous), leaving:

$$\left[\frac{\partial\tilde{\Phi}}{\partial z}\right]_{-\varepsilon}^{\varepsilon} = 4\pi G\tilde{\Sigma}(k) \tag{4.54}$$

Compute the derivative:

- For $z > 0$, $\tilde{\Phi} = C(k)e^{-kz}$, so $\frac{\partial\tilde{\Phi}}{\partial z} = -kC(k)e^{-kz}$, and at $z = 0^+$, $\frac{\partial\tilde{\Phi}}{\partial z} = -kC(k)$.
- For $z < 0$, $\tilde{\Phi} = C(k)e^{kz}$, so $\frac{\partial\tilde{\Phi}}{\partial z} = kC(k)e^{kz}$, and at $z = 0^-$, $\frac{\partial\tilde{\Phi}}{\partial z} = kC(k)$.

Thus,

$$\left[\frac{\partial \tilde{\Phi}}{\partial z}\right]_{0^-}^{0^+} = -kC(k) - kC(k) = -2kC(k) \tag{4.55}$$

$$-2kC(k) = 4\pi G\tilde{\Sigma}(k) \quad \Longrightarrow \quad C(k) = -\frac{2\pi G\tilde{\Sigma}(k)}{k} \tag{4.56}$$

So,

$$\tilde{\Phi}(k, z) = -\frac{2\pi G\tilde{\Sigma}(k)}{k}e^{-k|z|} \tag{4.57}$$

Using the inverse Hankel transform:

$$\Phi(R, z) = \int_0^\infty k\tilde{\Phi}(k, z)J_0(kR)\, dk = \int_0^\infty k\left(-\frac{2\pi G\tilde{\Sigma}(k)}{k}e^{-k|z|}\right)J_0(kR)\, dk \tag{4.58}$$

$$\Phi(R, z) = -2\pi G\int_0^\infty \tilde{\Sigma}(k)e^{-k|z|}J_0(kR)\, dk \tag{4.59}$$

In the midplane $z = 0$.

$$\Phi(R, 0) = -2\pi G\int_0^\infty \tilde{\Sigma}(k)J_0(kR)\, dk \tag{4.60}$$

The circular velocity is,

$$V_d^2(R) = R\frac{\partial \Phi}{\partial R}\bigg|_{z=0} \tag{4.61}$$

Differentiate $\Phi(R, z)$:

$$\frac{\partial \Phi}{\partial R} = \frac{\partial}{\partial R}\left(-2\pi G\int_0^\infty \tilde{\Sigma}(k)e^{-k|z|}J_0(kR)\, dk\right) \tag{4.62}$$

Since $\tilde{\Sigma}(k)$ is independent of R, we move the derivative inside the integral

$$\frac{\partial \Phi}{\partial R} = -2\pi G\int_0^\infty \tilde{\Sigma}(k)e^{-k|z|}\frac{\partial}{\partial R}J_0(kR)\, dk \tag{4.63}$$

Using $\frac{dJ_0(x)}{dx} = -J_1(x)$, so $\frac{\partial J_0(kR)}{\partial R} = -kJ_1(kR)$.

$$\frac{\partial \Phi}{\partial R} = -2\pi G\int_0^\infty \tilde{\Sigma}(k)e^{-k|z|}(-kJ_1(kR))\, dk = 2\pi G\int_0^\infty k\tilde{\Sigma}(k)e^{-k|z|}J_1(kR)\, dk \tag{4.64}$$

At $z = 0$:

$$\frac{\partial \Phi}{\partial R}\bigg|_{z=0} = 2\pi G\int_0^\infty k\tilde{\Sigma}(k)J_1(kR)\, dk \tag{4.65}$$

Thus,

$$V_d^2(R) = R \cdot 2\pi G \int_0^\infty k\tilde{\Sigma}(k) J_1(kR)\, dk = 2\pi G R \int_0^\infty k\tilde{\Sigma}(k) J_1(kR)\, dk \tag{4.66}$$

For $\Sigma(R) = \Sigma_{d0} e^{-R/R_d}$,

$$\tilde{\Sigma}(k) = \int_0^\infty R\Sigma_{d0} e^{-R/R_d} J_0(kR)\, dR \tag{4.67}$$

Substitute $u = R/R_d$, so $R = R_d u$, $dR = R_d\, du$:

$$\tilde{\Sigma}(k) = \Sigma_{d0} \int_0^\infty (R_d u) e^{-u} J_0(kR_d u) R_d\, du = \Sigma_{d0} R_d^2 \int_0^\infty u e^{-u} J_0(kR_d u)\, du \tag{4.68}$$

This is a standard integral: $\int_0^\infty x e^{-ax} J_0(bx)\, dx = \frac{a}{(a^2 + b^2)^{3/2}}$, with $a = 1$, $b = kR_d$:

$$\int_0^\infty u e^{-u} J_0(kR_d u)\, du = \frac{1}{(1 + (kR_d)^2)^{3/2}} \tag{4.69}$$

So,

$$\tilde{\Sigma}(k) = \Sigma_{d0} R_d^2 \frac{1}{(1 + (kR_d)^2)^{3/2}} \tag{4.70}$$

Substitute $\tilde{\Sigma}(k)$ into the velocity expression,

$$V_d^2(R) = 2\pi G R \int_0^\infty k \left(\frac{\Sigma_{d0} R_d^2}{(1 + (kR_d)^2)^{3/2}} \right) J_1(kR)\, dk \tag{4.71}$$

$$V_d^2(R) = 2\pi G \Sigma_{d0} R_d^2 R \int_0^\infty \frac{k J_1(kR)}{(1 + (kR_d)^2)^{3/2}}\, dk \tag{4.72}$$

Make the substitution $u = kR_d$, so $k = u/R_d$, $dk = du/R_d$,

$$\begin{aligned} V_d^2(R) &= 2\pi G \Sigma_{d0} R_d^2 R \int_0^\infty \frac{(u/R_d) J_1((u/R_d)R)}{(1 + u^2)^{3/2}} \frac{du}{R_d} \\ &= 2\pi G \Sigma_{d0} R_d^2 R \cdot \frac{1}{R_d^2} \int_0^\infty \frac{u J_1(uR/R_d)}{(1 + u^2)^{3/2}}\, du \\ &= 2\pi G \Sigma_{d0} R \int_0^\infty \frac{u J_1(uR/R_d)}{(1 + u^2)^{3/2}}\, du \end{aligned} \tag{4.73}$$

Let $\xi = R/R_d$, so the integral becomes:

$$\int_0^\infty \frac{u J_1(\xi u)}{(1 + u^2)^{3/2}}\, du \tag{4.74}$$

Let $y = R/(2R_d) = \xi/2$, so $\xi = 2y$, and the argument $\xi u = 2yu$. The integral matches a known form (Binney & Tremaine 2008):

$$\int_0^\infty \frac{uJ_1(2yu)}{(1+u^2)^{3/2}}\,du = y[I_0(y)K_0(y) - I_1(y)K_1(y)] \tag{4.75}$$

where I_n and K_n are modified Bessel functions of the first and second kinds. Thus,

$$V_d^2(R) = 2\pi G\Sigma_{d0}R \cdot y[I_0(y)K_0(y) - I_1(y)K_1(y)] \tag{4.76}$$

Substitute $R = 2R_d y$,

$$\begin{aligned} V_d^2(R) &= 2\pi G\Sigma_{d0}(2R_d y) \cdot y[I_0(y)K_0(y) - I_1(y)K_1(y)] \\ &= 4\pi G\Sigma_{d0}R_d y^2[I_0(y)K_0(y) - I_1(y)K_1(y)] \end{aligned} \tag{4.77}$$

Hence,

$$V_d^2(R) = 4\pi G\Sigma_{d0}R_d y^2[I_0(y)K_0(y) - I_1(y)K_1(y)] \tag{4.78}$$

where:

- $y = \frac{R}{2R_d}$,
- Σ_{d0}: Central surface density,
- R_d: Exponential scale length,
- G: Gravitational constant,
- $I_n(y)$, $K_n(y)$: Modified Bessel functions.

Alternatively, express it explicitly in terms of R,

$$\begin{aligned} V_d^2(R) &= 4\pi G\Sigma_{d0}R_d\left(\frac{R}{2R_d}\right)^2\left[I_0\left(\frac{R}{2R_d}\right)K_0\left(\frac{R}{2R_d}\right) - I_1\left(\frac{R}{2R_d}\right)K_1\left(\frac{R}{2R_d}\right)\right] \\ &= \frac{\pi G\Sigma_{d0}R^2}{R_d}\left[I_0\left(\frac{R}{2R_d}\right)K_0\left(\frac{R}{2R_d}\right) - I_1\left(\frac{R}{2R_d}\right)K_1\left(\frac{R}{2R_d}\right)\right] \end{aligned} \tag{4.79}$$

Thus, the circular velocity is

$$V_d(R) = \sqrt{4\pi G\Sigma_{d0}R_d y^2[I_0(y)K_0(y) - I_1(y)K_1(y)]} \tag{4.80}$$

4.8 Appendix B: Tutorials

This chapter has accompanying supplementary material. The following Jupyter notebook tutorials

- Rotation curve from extreme velocities.
- Components of the rotation curve.
- Galactic structure from observations.
- LAB survey tutorial.

The latest versions of the tutorials can be found in: https://github.com/Ashish-Mhaske/BHARAT_Tutorials

References

Binney, J., & Tremaine, S. 2008, Galactic Dynamics, (2nd edn.; Princeton, NJ: Princeton Univ. Press)
Bowman, J., Rogers, A., & Monsalve, R. 2018, Natur, 555, 67
Burton, W. B. 1971, A&A, 10, 76
Christiansen, W. N., & Hindman, J. V. 1952, AuSRA, 5, 437
Clemens, D. P. 1985, ApJ, 295, 422
de Lera Acedo, E., de Villiers, D. I. L., & Razavi-Ghods, N. 2022, NatAs, 6, 984
Dickey, J. M., & Lockman, F. J. 1990, ARA&A, 28, 215
Dickey, J. M., McClure-Griffiths, N. M., Gaensler, B. M., & Green, A. J. 2003, ApJ, 585, 801
Ewen, H. I., & Purcell, E. M. 1951, Natur, 168, 356
Freeman, K. C. 1970, ApJ, 160, 811
Heiles, C., & Troland, T. H. 2003, ApJS, 145, 329
Heiles, C., & Troland, T. H. 2004, ApJS, 151, 271
Kalberla, P. M. W., & Kerp, J. 2009, ARA&A, 47, 27
Kalberla, P. M. W., Burton, W. B., Hartmann, D., et al. 2005, A&A, 440, 775
Kormendy, J., & Kennicutt, R. C. 2004, ARA&A, 42, 603
Kwee, K. K., Muller, C. A., & Westerhout, G. 1954, BAN, 12, 211
Lin, C. C., & Shu, F. H. 1964, ApJ, 140, 646
Mhaske, A. A., Bagchi, J., Joshi, B. C., Jacob, J., & Paul, K. T. 2022, AmJPh, 90, 948
Muller, C. A., & Oort, J. H. 1951, Natur, 168, 357
Navarro, J. F., Frenk, C. S., & White, S. D. M. 1996, ApJ, 462, 563
Oort, J. H. 1927, BAN, 3, 275
Oort, J. H. 1932, BAN, 6, 249
Oort, J. H. 1940, ApJ, 91, 273
Oort, J. H., Kerr, F. J., & Westerhout, G. 1958, MNRAS, 118, 379
Rubin, V. C., Ford, W. K., & Thonnard, N. 1980, ApJ, 238, 471
Sellwood, J. A., & Sparke, L. S. 1988, MNRAS, 231, 25P
Shu, F. H. 1970, ApJ, 160, 99
Simonson, S. C. 1976, A&A, 46, 261
Singh, S., Subrahmanyan, R., Shankar, N. U., et al. 2018, ExA, 45, 269
Sofue, Y. 2015, PASJ, 67, 75
Sofue, Y. 2020, Galax, 8, 37
Sofue, Y., & Rubin, V. 2001, ARA&A, 39, 137
Toomre, A., & Toomre, J. 1972, ApJ, 178, 623
Van De Hulst, H. C. V. 1982, Origin of the Radio Waves from Space (Netherlands: Springer) 302
Wakker, B. P., & van Woerden, H. 1997, ARA&A, 35, 217
Walter, F., Brinks, E., de Blok, W. J. G., et al. 2008, AJ, 136, 2563
Zwicky, F. 1937, ApJ, 86, 217

Part II

Local Stations That Are Part of Larger Network or Project

Chapter 5

Hydrogen Map of Our Milky Way

Glen I. Langston and Kevin Bandura

There are many amazing research projects carried out in the context of back-yard HI astronomy. Here we present the steps for collecting a few weeks of observations, so you can make a detailed, and beautiful, map of hydrogen gas Milky Way. This project will give students a Galactic perspective. The goal of this project is to convince students that we are in the Milky Way's disk of stars and are surrounded by the hydrogen gas that is forming new stars. *Estimated Cost of DSPIRA feedhorn: several $100s USD*

5.1 Introduction

The project involves setting a horn radio telescope outside for two or three weeks, to observe all parts of the sky visible at your location. The software used here is specific to the Digital Signal Processing in Radio Astronomy (DSPIRA) feedhorn design developed at West Virginia University.[1] Most of this material could also be done with the BHARAT feedhorn design from Chapter 3. It is more productive to have several students build telescopes, learning to work together and more quickly combining their observations to discover the Milky Way always visible to them, day or night. Your completed project will yield a beautiful map of the sky, like that in Figure 5.1.

The figure shows the whole northern sky, visible from the author's back yard. The map shows the total amount of hydrogen on a map of sky coordinates. The X-axis is *Right Ascension (RA)*, from 24 hours on the left to 0 hours on the right. The sky, and this image, are connected at the edges, so that RA = 0 and RA = 24 are the same part of the sky. Every day, every Right Ascension passes straight overhead, so if you point a telescope straight up, your telescope will see a complete stripe of the

[1] https://wvurail.org/dspira-lessons/

doi:10.1088/2514-3433/ae1223ch5

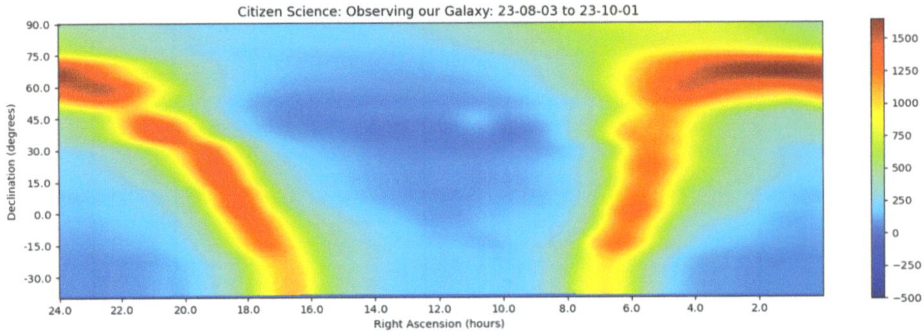

Figure 5.1. Right ascension, declination map of the entire sky visible from Latitude 38.4 degrees. The bright red strip curving its way across the northern sky is part of the disk of the Milky Way. Since we are in the disk of the Milky Way, it shows up a stripe of strong emission surrounding us. We can see the center of the Milky Way, but not all of the Milky Way from northern latitudes. Between 6 and 16 hr Right Ascension the Milky Way disk is below our horizon.

sky in 24 hr. The Y-axis shows the astronomical *Declination*. You will observe different Declinations by pointing your telescope north or south. Astronomers at different latitudes will see slightly different regions of Declination, for the same telescope elevation.

Your horn will look something like one of the two horns shown in Figure 5.2.

5.2 Notes from Previous Chapters

The process of making a map of the sky is fairly simple, after you have completed building a telescope. It is more fun to do your mapping project as a group research project, if you and several others work together to build telescopes, then you can share making parts of the observations. Working as a group will also help in learning how to make telescopes better, by comparing different telescopes. As a group, you can also learn cooperation and leadership, then plan new experiments. Modern astronomy is carried out by groups of researchers.

Descriptions of how to build a feedhorn telescope can be found in Chapter 3, using the example of Bose Horn Antenna Radio Telescope (BHARAT). The work in the present chapter is based on a similar but distinct classroom feedhorn design called Digital Signal Processing in Radio Astronomy (DSPIRA). Additional lessons can be found here:

https://wvurail.org/dspira-lessons/

In particular, the "Lightwork" series of memos and videos here provide good instructions for building a DSPIRA base and antenna, in particular Memos 35 and 36:

https://wvurail.org/lightwork/

In this chapter, you'll be making a *total intensity* map of hydrogen in the Milky Way.

Figure 5.2. The neutral hydrogen telescope is a "Pair of Pails" telescope. This telescope, made from two pails, stacked on top of each other. The bigger pail is 13 inches wide at the top, and the bottom is cut out to allow gathering the signal in the bottom pail. The Pails are sitting on the simplest horn telescope base. For scale, Oni, our dog is about 30 inches tall when sitting. The concrete blocks hold the horns in place, in case of high wind.

5.3 Mapping Software Setup

We describe the steps involved in making a map of the Milky Way. All software used for this project is written in `Python`. The GNURadio[2] software system generates Python code that runs on the Raspberry pi to make the observations. The `analyze` programs are also written in Python and can run on the Pi computer or any other Linux computer that has access to the saved observations.

The mapping observations can be made using several different Software Defined Radios (SDRs). As a good compromise between higher sensitivity and lower cost,

[2] https:://www.gnuradio.org

we're assuming you are using the Airspy Mini SDR.[3] This SDR operates well at 1421.5 MHz and has a 6 MHz bandwidth. The amplifier gain chain in the SDR is somewhat complicated, but for our purposes set the first gain value to 21 dB and the other two gains to 0. Using the interface, turn the bias tee on, and add your name and the name of your telescope.

If using a different SDR, there is software support for SDRPlay RSP1A, Pluto SDR, and NooElec SDRs. There are different computer programs for each device. The Gnuradio-based program, called `NsfIntegrate60`, is the program that controls the Airspy. All the real-time software for the Raspberry Pi is on the WVURAIL Github site:

www.github.com/WVURAIL/gr-radio_astro

All the Python software for analyzing the saved observations is downloaded from the author's `analyze` Github site:

www.github.com/glangsto/analyze

The Python programs used to create the images are described in the next sections.

5.4 First Calibration

After you have set up your telescope and computer for observations, use the `NsfIntegrate60` program to start observations. Set up the telescope for a 6 MHz bandwidth observation, with center frequency of 1421.5 MHz. Confirm, when you point your telescope straight down, that the program shows a strong signal. The average level should be between 2 and 6 across most of the plot. Set the telescope azimuth to 0, point the telescope base toward north and set the telescope elevation to −90 degrees. Click "Survey" on the interface, then click "Wait" to clear any data previously averaged. Next click "Average." You are observing the "hot" ground for a few minutes (3 or more). Click the "Hot/Cold" box on the user interface and then click the "Save" box. This saves a file called `Hot.hot` in the current directory. After a few seconds, click the "Survey" and then select "Wait". Figure 5.3 shows the GNURadio created user interface program with a plot of the hot and cold load observations.

The first hot-load observation is done. Go outside, align the telescope base with due north again and put the horn on the mount. Put the computer box under the telescope. It is important to put some sort of metal, like aluminum foil, or a mylar bag over the computer, to protect the observations from interference created by the computers. Also keep your phones and other electronics away from the telescopes.

Back inside, set the telescope elevation to +90. The signal level should have dropped significantly, by a factor of 2 or 3 compared to the hot load. Start the observation by clicking the "Average" button. Observe for about the same time as for your hot-load observation. Click the "hot/cold" button, then click the "Save" button. This saves a file called `Cold.ast` in the current directory. You've

[3] https://airspy.com

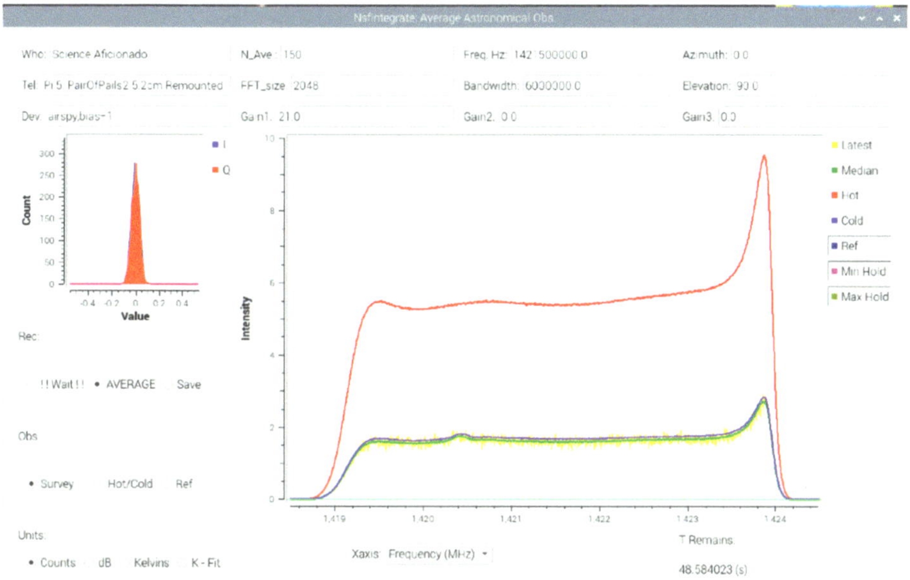

Figure 5.3. The `NsfIntegrate60` program, running on the Raspberry Pi computer, showing the user inputs and a plot of raw counts. Four curves are shown in the plot of Intensity on the Y-axis versus Frequency (MHz) on the X-axis. The upper, red, plot shows the intensity measured when the horn is pointed straight down, elevation $= -90$. There are three lower curves, showing intensity in when the horn is pointed straight up. The hydrogen in our Galaxy is seen as a weak bump in the lower plots at 1420.4 MHz.

completed your first test! Figure 5.6 (left) shows the comparison of hot and cold load spectra. On the right side of the figure is a zoom in on the calibrated intensity.

Click the "Kelvin" button. The plot will change to show Intensity, in Kelvins, on the Y-axis. The X-axis will not change. See Figure 5.4. The calibration of the sky observation is computed very quickly. The intensity is measured in "Kelvins" a unit of temperature. You may have to zoom out, with the mouse, to find the calibrated spectrum. If the average temperature you measure is less than 250 K, you can start the mapping observations. If the temperature you measure is higher, there maybe a loose connection somewhere in the coaxial cables. Tighten all the cable connections and try the calibration again.

When the measured temperature is low enough, but before we map the sky, we should make a deep calibration observation, observing the ground for 4 hours. As you did before, at the end of your long observation save the data, which replaces the file `Hot.hot` with the latest version. In a notebook, record the average system temperature you measure. The lower the system temperature you measure, the better.

The interface program can also show the current plot of intensity, from only a two second average. A baseline is subtracted and the spectra are smoothed and filter to reduce interference. Choose the "K-Fit" option. The resulting plot should look similar to Figure 5.5.

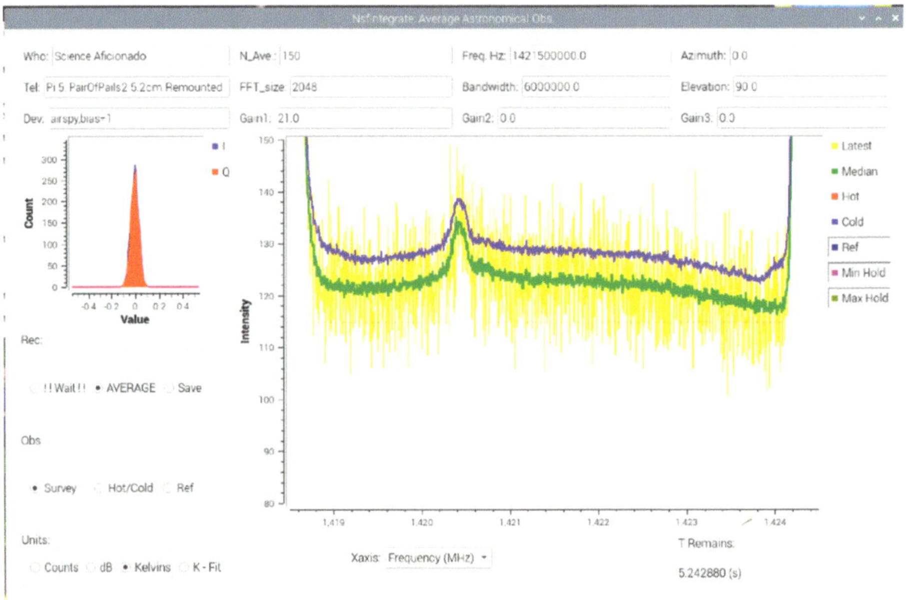

Figure 5.4. The `NsfIntegrate60` program again, with the plot showing the calibrated signal, measured in Kelvins. Three curves show Intensity on the Y-axis versus Frequency (MHz) on the X-axis. The upper, blue, plot shows the intensity in Kelvins, based on the first cold load average intensity. The lower, green, plot shows the most recent average intensity measured. **After calibration, the hydrogen emission is much more clearly seen in the blue and green plots, as peaks in the plot at 1420.4 MHz.** The two curves are separated due to gain changes in the amplifiers and some small change in the brightness of the sky. The yellow curve shows the intensity, in Kelvins, measured in a single, very short, 0.00034 s average.

5.5 Sensitive Calibration

The sensitive calibration consists of two parts, a long hot-load observation and a high elevation observation that we'll save for use in calibrating all the other mapping observations.

We need to make a long hot-load observation, to reduce the average noise contributed by the hot-load calibration. Unbolt your horn from the base and point it at the ground again, without disconnecting, or stressing, the coaxial cable connecting the amplifier to the computer.

Use the program `NsfIntegrate60` to indicate the telescope is pointing down (Elevation = −90). Click the "Wait" button to clear any data averaged before the horn was pointed down, then click "Average." The observations will start, and data will be written to the Raspberry pi computer every few minutes.

After 4 hours have passed, compare the latest average to your previously saved hot-load observation. There should only be a few % change (less than 10%) in the intensity compared to the first, short hot-load measurement.

If the two spectra are similar, save your new hot-load spectrum. To do this click "Hot/cold," then click "Save." Next we'll prepare for our first full day of observations of the sky. Click "Survey," then "Wait" on the interface. Go back

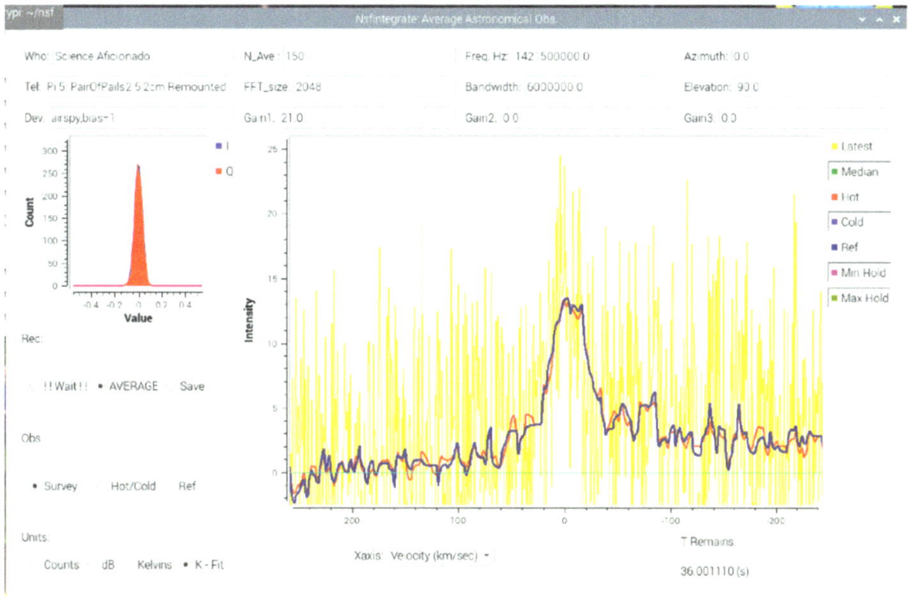

Figure 5.5. The `NsfIntegrate60` program with the plot zoomed in on the intensity of the peak emission. In this plot of Intensity in Kelvins, on the Y-axis, and Velocity in km s^{-1}, on the X-axis. The purple curve shows the latest 2 s average intensity, after subtracting a baseline. The red curve shows the previous 2 s average. Note that the Milky Way is clearly visible, with this very small radio telescope, in just 2 s. The yellow curve shows the intensity, in Kelvins, measured in a single, very short, 0.00034 s average.

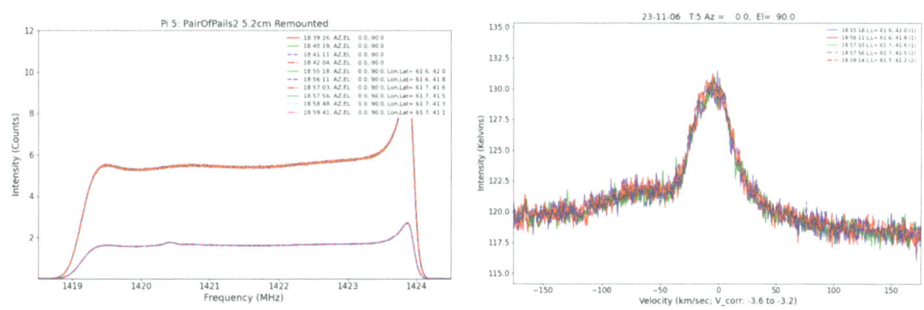

Figure 5.6. Check of the saved first hot/cold load observations. The figure on the left shows intensity in raw counts versus frequency in MHz. The upper curves show the hot load and lower curves show the sky. Note, again, that the ground is much hotter than the sky. Also note very good agreement between the separate hot-load observations and cold load observations in a minutes. The figure on the right shows the calibrated intensity based on the raw observations. Velocity (km s^{-1}) is on the X-axis, and Intensity (in Kelvins) is on the Y-axis. The average system temperature is around 119 K. The local arm of the Milky Way is clearly visible as the peak near $V = 0$. **Note the weak feature at about -75 km s^{-1}, which hints we see another spiral arm of the Milky Way.**

outside, put the horn back on the base and point the telescope straight up. Double check that the base is still aligned with North.

Go inside, set the elevation to +90 and click "Average." We'll wait one whole day for the observation to compete.

5.6 `Analyze` Spectra: Use `R` and `T` Commands

While the telescope is observing, you can look at the previously saved observations. The data files are being saved on the Raspberry Pi computer.

If the weather is good and no leaves fall in the horn, then usually all the spectra good and can be used for a map. It will save you some time if you first check all your spectra before preceding to the mapping stage. These are the steps for checking your calibration and spectra:

- Open another terminal window on the Raspberry Pi.
- Go to the data directory using Linux command:

  ```
  cd ~ /data
  ```

- Create directories for each day of observation and move the data to that directory. Each Raspberry Pi is assigned a number, in this case 5 (when using five telescopes on site). Here use date 24 October 23, abbreviated as 23Oct24:

  ```
  mkdir ~/pi5-data-23Oct24

  mv 23-10-24T* ~/pi5-data-23Oct24

  cd ~/pi5-data-23Oct24
  ```

- Check the short (few minute duration) hot and cold spectra. Compare the intensity changes over a 10 hour period the `R` command. The few minutes of observations are broken up into different files. Use the `ls` command. Then find the first files you created. You can list or plot a few file using the Linux wild card "*": `ls *`

  Note that the hot-load files all end with extension `.hot` and the astronomy files end with extension `.ast`. Also note that the "`T`" in the file names separates the date and the time. To make sure you are selecting the hours part of the time, include the letter "T" in the Linux wild card. For example, to look at the observations taken during hour 18 UTC use: `R *T18*`.
- Look at your first few hot and cold observations. Compare the hot and cold load observations recorded. In the example here, the hot-load observations were taken at 18 hr 40 min UTC and the cold load observations were taken at 18 hr and 50 min.

  ```
  R *T18*.hot *T185*.ast
  ```

- Look at the first 10 hours of observations using `R *T0??0*` which plots all the data for 0 to 9 hours of the day, only plotting the spectra saved on the even minute. This reduces the plot to only showing 1/10 of all the data. See Figure 5.7.
- Look at the next 10 hours of data, using `R *T1??0*`. See Figure 5.8.
- Look at the final 4 hours of data using `R *T2??0*`
- Check the system temperature, using the short hot and cold spectra. Use the `T` command. Look at just the calibration from the first hot and cold load observations.

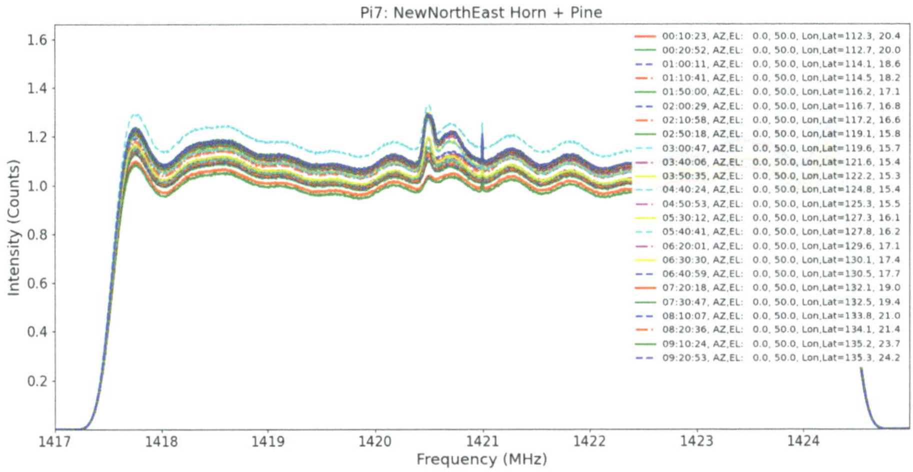

Figure 5.7. Check of the Raw spectra in the time range 0 to 9 hr. The X-axis is frequency and the Y-axis is intensity. The curves show the intensity in raw counts. **The hydrogen we are interested in is the peak near 1421.4 MHz.** Also note that when the hydrogen is stronger, the entire curve is a little higher. The Milky Way Galactic Plane also includes supernovae remnants which are radio bright, increasing the average height of the curve. Over the 10 hours, the amplifier gains change by about 10%, which also causes the plotted lines to move up and down. The amplifier changes limit the accuracy of your map intensities.

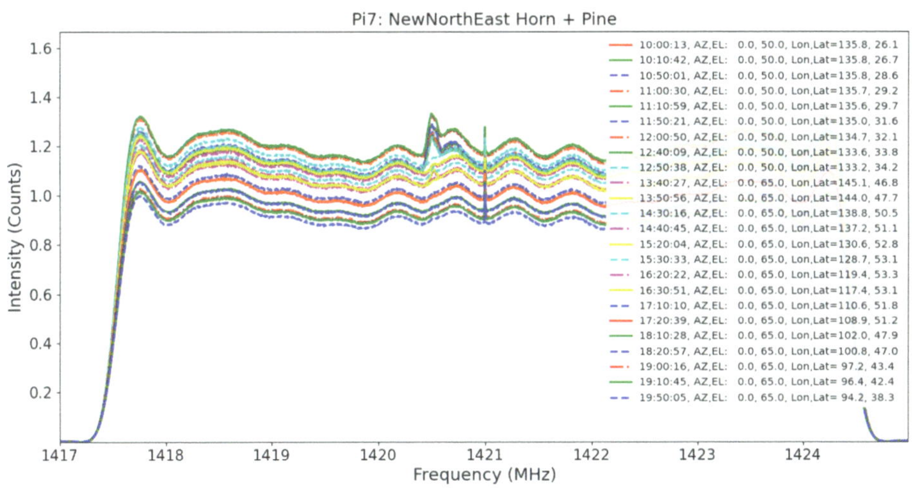

Figure 5.8. Check of the Raw spectra in the time range 10–19 hr. The axes are the same as in the previous figure.

Now start the main observations; point the telescope straight up, start observations with `NsfIntegrate60` and leave the system running for one whole day.

Since the data are relatively easy to obtain, simply delete any suspect data with unusually high system temperature, due to bad weather or leaves accumulating in the horn.

We will use these observations to create the reference spectra for calibration for all the remaining spectra.

5.7 Reference Hot and Cold Load: `T -K` Command

The `T` command has slowly been updated to include more and more features as new issues were found that had to be addressed. The program has a built-in documentation, that is listed when the `T` command is run without arguments. A few of the most used options are listed below:

- `-C` Flag the center channel in the spectra, that are sometimes corrupted.
- `-F` Flag known RFI lines, the problem frequencies are built into the program.
- `-B` Fit and remove a baseline from the spectra.
- `-K` Keep the computed hot and cold loads average spectra for later use.
- `-S` Summarize the observations, computing the peak and integrated intensity. This summary is used in the mapping step to compute the total intensity images.

The average hot and cold spectra are very important for making a quality image of the Milky Way. Make sure you've checked all the spectra for calibration. The process is fairly simple. This step does use a fairly large set of options to the `T` command. The first step is to create a directory for keeping the reference spectra. Because some collaborations use several telescopes, create a keep directory for each telescope. Here we use names like `pi5` and include a date for the period of observations, i.e.,:

```
mkdir ~/pi5-23Oct
```

Now compute and save the average hot and cold spectra.

```
T -K ~/pi5-23Oct -B 3600 *
```

The program will print the names of the hot and cold files kept. We'll need these calibration files in the next step. This program will also plot hourly averages spectra.

5.8 Summarizing the Observations: `T -S` Command

After a day of observations is completed, and you've checked that there were no spectra that are discrepant, then the first step is computing the integrated intensities for placing on the map. The integrated intensities are written to an ASCII log file. The log file contains time-tagged sums of the hydrogen intensity in the directions viewed by the telescopes.

For plotting, one can show plots of one hour averages, but for mapping we need to measure each spectrum separately. We will also want to subtract a baseline.

To create the file for input for mapping use the T program again, but with these arguments:

```
T -B -A <hot file> <cold file> -S pi5-23Oct.sum 60 *.ast
```

The hot and cold file names should be those from the previous step. The `-B` argument again means subtract a baseline. The `60` argument means average for 60 s. The `-S pi5-23Oct.sum` argument is the output summary file.

The program will plot only 24 files, but will process all of them, summarizing the intensity measurements (Figure 5.9). The intensities as a function of sky position can be accumulated and plotted as shown in Figures 5.10 and 5.11.

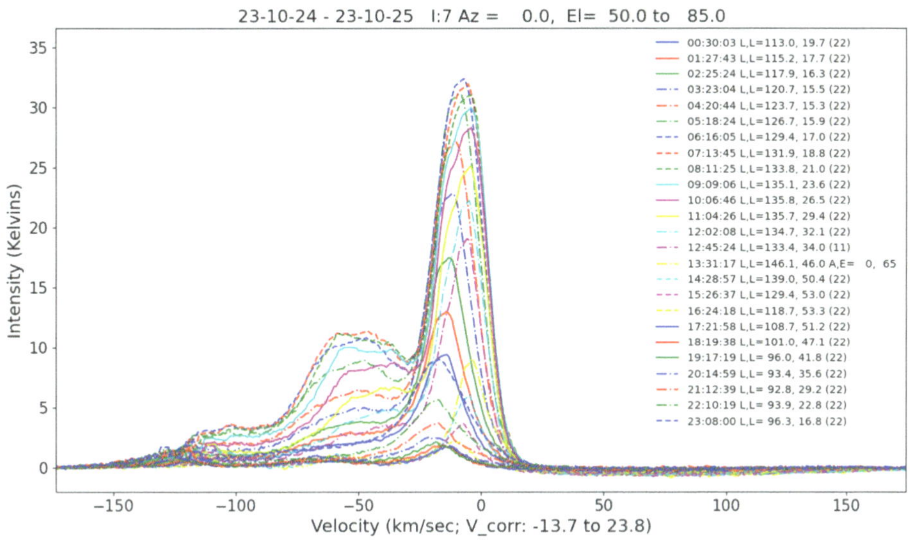

Figure 5.9. Plot of the calibrated intensities, showing the change in intensity of the Milky Way emission as a function of time. The X-axis is now velocity (km s^{-1}) over the range where the Milky Way is expected to be bright. The Y-axis intensity is in Kelvins, based on the hot/cold load calibration. The individual lines are one hour averages. The thicker lines mark spectra taken within 10 degrees of the Galactic plane. Note that the hydrogen emission is always visible, but is about 10 times stronger in the Galactic plane.

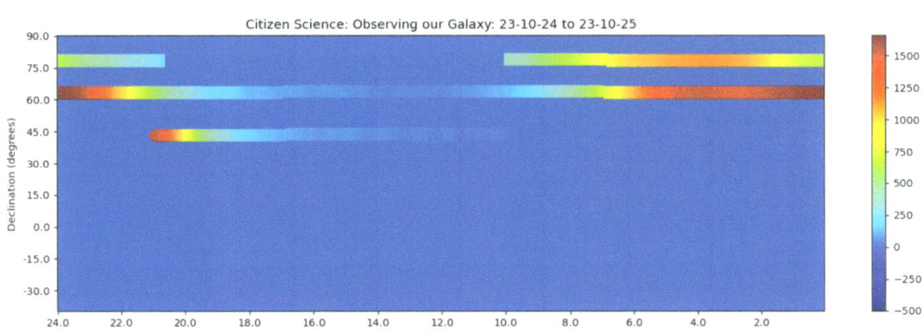

Figure 5.10. Right ascension, declination map of the entire sky visible from Latitude 38 degrees. The three horizontal stripes correspond to the different elevations observed during the days of October 24 and 25. The telescope was pointing North, Azimuth= 0, and the elevations were 50, 65, and 85 degrees. These elevations, at the latitude of our telescope, 38.4 degrees, correspond to roughly 78.4, 63.4, and 43.4 degrees declination. Note that each day has parts of two elevation observations.

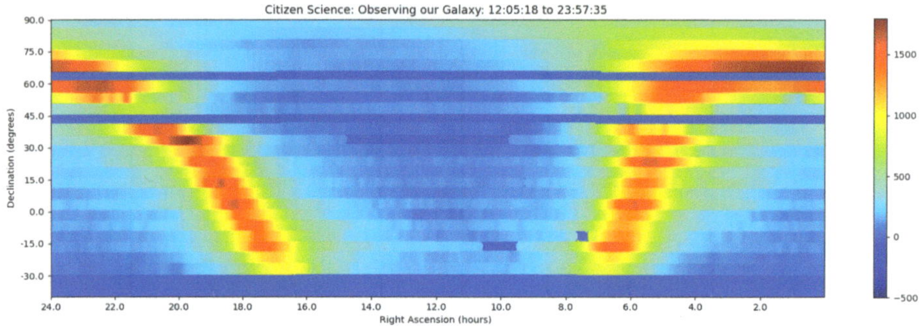

Figure 5.11. Similar to Figure 5.1, but showing the individual stripes.

Reference

The LightWork Memos are all available on the DSPIRA GitHub repository: https://github.com/WVURAIL/lightwork

AAS | IOP Astronomy

Radio Telescope Instrumentation for Teaching

Timothy Dolch

Chapter 6

The Radio JOVE Project

Chuck Higgins and Dave Typinski

Radio JOVE (radiojove.gsfc.nasa.gov) is a longstanding citizen science and public outreach project that uses radio telescopes operating between 15–30 MHz (the upper half of the high frequency or HF band) to observe solar and planetary radio emissions along with galactic and extra-galactic radio sources. We promote and train radio enthusiasts to add skills and knowledge to become high-level citizen scientists. We developed a low-cost ($400 USD) radio spectrograph instrument (16–24 MHz) and calibration system ($300 USD) that can address more science questions related to heliophysics, planetary and space weather science, and radio wave propagation. *Estimated Cost: $700 USD; this chapter also includes pure data analysis exercises not requiring equipment.*

6.1 Introduction

Figure 6.1 shows multiple solar radio bursts recorded with the Radio JOVE spectrograph system on 2021 July 9 in Murfreesboro, TN. The raw data are calibrated as antenna noise temperature, and the solar bursts are easily visible as enhancements above the galactic background, which is the dark blue color. Local interference is seen as horizontal lines/bands on the spectrograph, where a particularly strong interference band is seen at 17.5 MHz. The spectrograph captures scientific information about the radio source (Galaxy, solar, planetary, terrestrial), such as spectral enhancements, intensity variations, duration, spectral drift, and polarization if using a proper antenna.

6.1.1 Decameter Radio Wave Emissions

Radio waves sources include many human-made and natural radio emissions from Earth: radio stations, communications, lightning, and many types of radio interference. Radio astronomy sources of radio emission include solar system objects such as the Sun and planets, many types of objects within the Milky Way, extra-galactic sources like radio galaxies, and cosmological signals from the beginnings of the

doi:10.1088/2514-3433/ae1223ch6

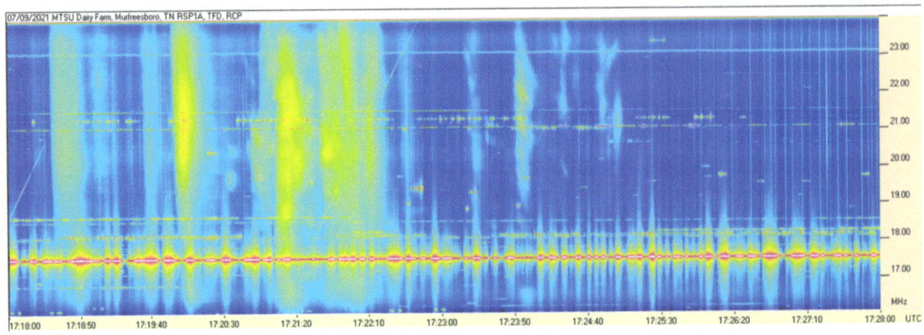

Figure 6.1. Multiple solar radio bursts recorded at 20.1 MHz on 2021 July 9, using an SDR radio spectrograph from 16 to 24 MHz. Local interference is seen as horizontal band in the spectrograph, particularly at 17.5 MHz.

Universe. Radio emission is generated in two categories, thermal emission due to the temperature of the object, and nonthermal emission from charged particle motion in local magnetic fields. Radio emissions in the HF radio band are primarily associated with nonthermal processes, seen in Figure 6.2, an outgoing solar wind shock, a radio image of Jupiter, radio jets from galaxy Centaurus A, and a map of radio emission from the Milky Way.

Nonthermal radio emission is a very important process, particularly gyromagnetic processes. Gyromagnetic emissions involve moving charged particles in magnetic fields like planetary or solar magnetic fields. Depending on the energy of the moving charge, cyclotron or synchrotron radio emission results. Pulsar radio emission is thought to result from a gyrosynchrotron process where a rotating neutron star emits radio waves from the plasma near the intense magnetic poles. Finally, radio emission is also generated by a microwave maser process, sources including molecular clouds and some planetary and solar magnetospheres.

6.1.2 Astronomical Targets Jupiter, the Sun, and the Galaxy

Jupiter's low-frequency radio emissions are strongly circularly polarized and are influenced by Jupiter's magnetic field and rotation, the solar wind, and the Io plasma torus. In particular, the most powerful Jovian emissions occur from nonthermal decameter (DAM) sources from 3 to 30 MHz and are the Io-DAM — decametric emission tied to Io flux tube and Io torus and Non-Io-DAM — decametric are related to auroral emissions. Jupiter decameter emission is a cyclotron maser mechanism process where electrons are accelerated by strong electric and magnetic fields and emit radio waves in a cone-like pattern. Jupiter's magnetic field magnitude limits the DAM emissions to below 39.5 MHz. However, higher energy synchrotron emissions in Jupiter's magnetosphere occur from 40 MHz to a few GHz, and higher frequencies are dominated by the thermal radio emission coming from the planet itself.

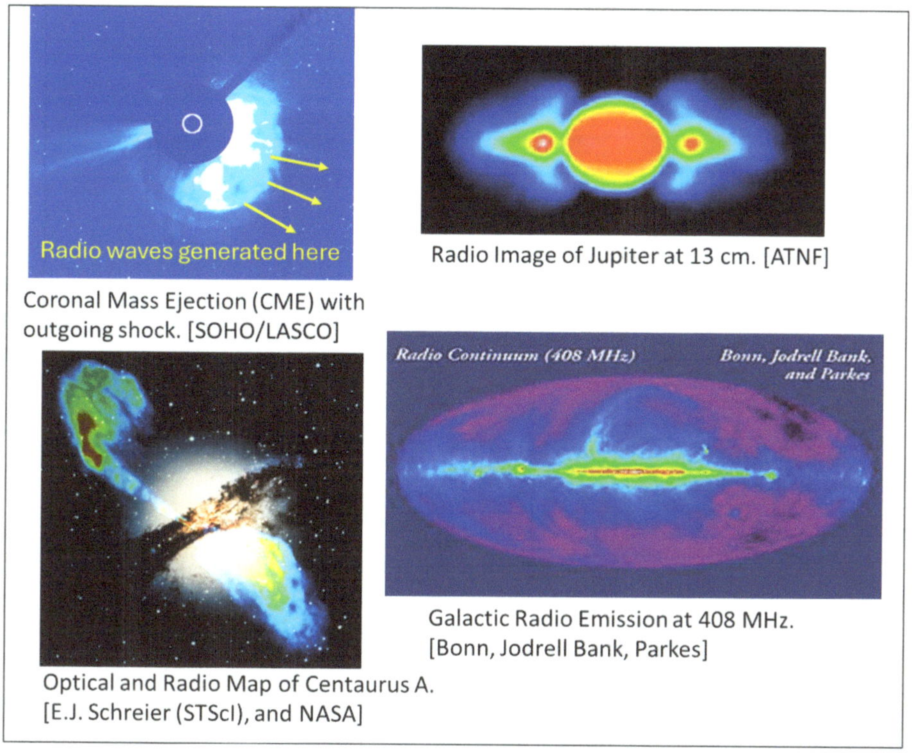

Figure 6.2. Four different sources of nonthermal radio emissions, the Sun, Jupiter, radio galaxies, and the Milky Way. Top left: Reproduced from (Gopalswamy et al., 2018). © 2018. The American Astronomical Society. CC BY 3.0. Bottom left: Science Release Credit: E.J. Schreier (STScI), and NASA. Top right: CSIRO Radio Astronomy Image Archive. Bottom right: Image credit NASA.

The Sun emits a broad spectrum of solar radio emissions, from a few kHz to hundreds of GHz from mechanisms that convert electron motions into electromagnetic radiation. Decameter (3–30 MHz) solar radio emissions are mostly unpredictable but are very common, especially around times of high sunspot number. Energy releases from solar flares and coronal mass ejections (CMEs) are the drivers of these nonthermal, gyromagnetic radio emission, called solar radio bursts. The bursts can be very strong and easily observed with a basic radio telescope like Radio JOVE. Solar radio emissions are generally unpolarized, thus a simple dipole antenna can be used to record them.

The Milky Way provides a lower limit to radio observations since the galactic emissions are pervasive as a result of spiraling electrons within the Galactic magnetic field. The Galactic background intensity can be measured and used as a baseline for other sources like Jupiter, solar, and even terrestrial signals. It is an important value to determine whether your local observing conditions are quiet or riddled with radio frequency interference (RFI).

6.2 The Radio JOVE Telescope

6.2.1 Basic Theory of Operation

6.2.1.1 Antenna

In simple terms, a radio antenna intercepts energy from electromagnetic waves and converts that energy into an electrical signal at the antenna terminals. This weak radio frequency signal is fed from the antenna to the receiver by a coaxial transmission line that minimizes losses and distortion (Radio JOVE 2012).

The radio antenna recommended for Radio JOVE is two half-wave dipole antennas (Figures 6.3 and 6.4), requiring about 30 feet × 40 feet (9 m × 12 m) of space. A single dipole is the minimum needed to detect solar bursts and the strongest Jupiter emissions. Two dipoles are required to detect weaker Jupiter and solar emissions. Hardware, coaxial cables, antenna wire, masts, guy ropes, and stakes are easily obtained from a hardware store. See the current antenna construction manual, the Radio JOVE Antenna Manual (2025): https://radiojove.gsfc.nasa.gov/radio_-telescope/manuals/RJ_Antenna_Manual_v2.1_2025_02_07.pdf.

Due to increasing ionospheric absorption "cutting off" cosmic signals below 15 MHz, and due to the Jupiter emission signal strength decreasing above 24 MHz,

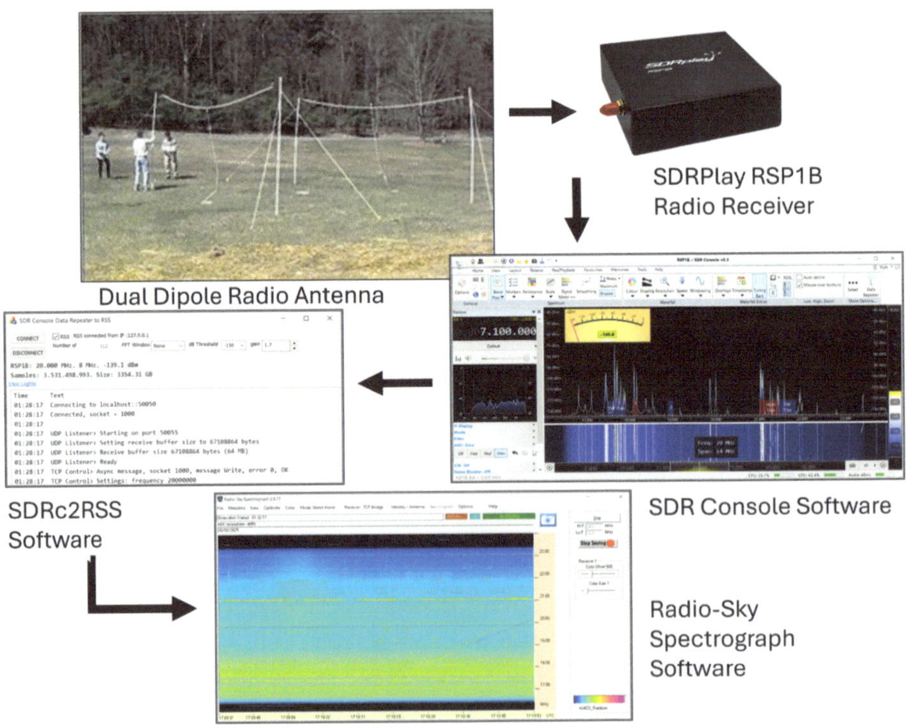

Figure 6.3. Radio JOVE Hardware: Dual Dipole antenna, SDRplay RSP1B receiver, SDR Console and SDRc2RSS radio control software, and Radio-Sky Spectrograph (RSS) software. Note that kit does not include antenna support structure. SDRplay (www.sdrplay.com) is a UK-based company that manufactures Software Defined Radio (SDR) radios. Radio-Sky Spectrograph software from radiosky.com.

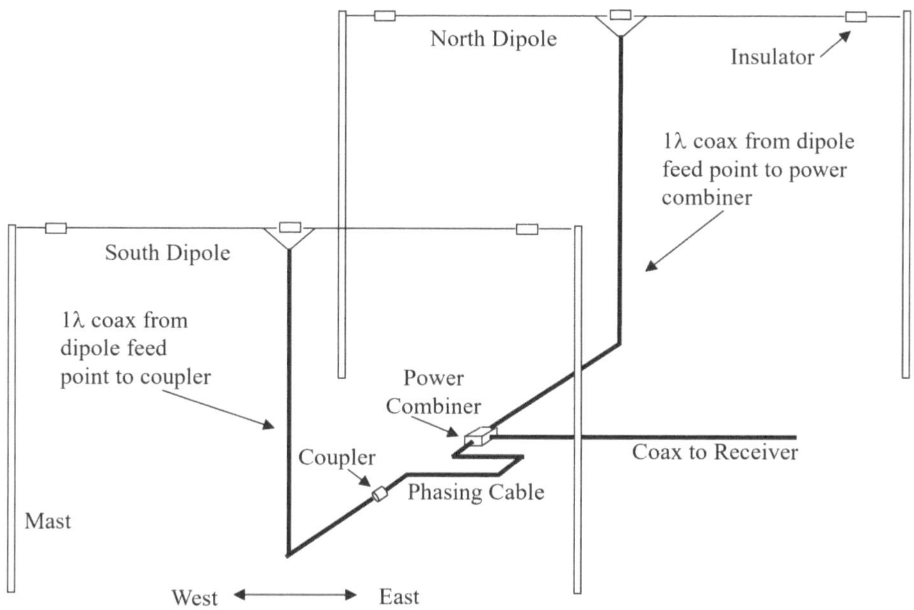

Figure 6.4. The Radio JOVE dual dipole antenna with a phasing cable inserted in the transmission line leading from the south dipole to the power combiner. This is a configuration for northern hemisphere observers. For an observer in the southern hemisphere the phasing cable would be installed in the line leading from the north dipole to the power combiner. Reproduced from JOVE (2012).

the optimal frequency range for ground-based Jupiter observations is from 18 to 22 MHz. Therefore, antennas designed for a center frequency near 20 MHz (wavelength = 15 m) are ideal for Jupiter and, beneficially, also work very well to observe solar radio emissions.

For a 20 MHz dipole, the tip-to-tip length is 23.4 feet (7.1 m). To achieve an antenna beam pattern (i.e., maximum gain) pointing upward, the dipole is mounted horizontally a quarter wavelength ($\lambda/4$) above the ground. The added benefit of a horizontal dipole with an upward-pointing beam is that it is less sensitive to terrestrial interference, most of which comes from the horizon. The $\lambda/4$ height is optimal to account for reflected waves off the ground that are combined in phase (constructive interference). The ground acts as a reflector for radio waves if the conductivity of the soil is good quality; a rich and moist soil is very good quality (Flagg 2005). Changing the height of the dipole antenna will change the pattern of the beam. For example, a dipole located a half wavelength ($\lambda/2$) above ground will cause a null overhead and a split beam 30° above the horizon (Flagg 2005).

To detect weaker radio emissions, Radio JOVE developed an antenna array using two dipole antennas (Figure 6.4) to achieve almost twice the gain of a single dipole. Without a phasing cable, the two antennas produce an overhead beam, somewhat narrower than that of a single dipole alone. It is an important principle of antennas that when the gain is higher the antenna beam size will be narrower. The power combiner adds the two signals that are in phase, resulting in the power of a point source signal sent to the receiver that is twice that of a single dipole alone (Radio JOVE 2012).

6.2.1.2 Receiver

Radio signals from astrophysical sources are typically very weak – for example, radio emissions from Jupiter produce less than a millionth of a volt (1 microvolt, 1 μV) at the antenna terminals of the receiver (Flagg 2005). These weak radio frequency (RF) signals must be amplified and filtered by the radio receiver, currently the SDRplay RSP1B (Figure 6.3). See SDRplay (https://www.sdrplay.com/) for more information about this versatile radio and its abilities. The SDR receiver, the accompanying control software and Radio-Sky Spectrograph (RSS) recording software, and the radio antenna are designed to operate over an 8 MHz (megahertz) bandwidth centered on 20 MHz. This frequency range is optimum for detecting Jupiter radio signals and is excellent for detecting solar radio emissions, the galactic background from the Milky Way, and natural and artificial terrestrial radio emissions. The advantage of wide-bandwidth observations (spectrograms) over narrow-band observations (strip charts) is that spectral information about the radio emitter leads to a much better understanding of the physics behind the radio wave generation and propagation.

6.2.1.3 Computer and Software

The computer and software will record and provide a real-time display of the strength of the digital radio signal on a frequency vs. time graph (a spectrogram). Radio JOVE requires a computer running Windows 7 or later operating system because the software-defined radio (SDR) requires a reasonable CPU speed to process the signals. Ideally, the computer should be dedicated to data collection when in operation, but it will depend on the capability of your computer.

The Radio-Sky Spectrograph[1] software allows you to view and store signals from software-defined radios (SDRs) or from special frequency-sweeping receivers. These radio telescopes, or "radio spectrographs", allow you to see how radio signals are distributed across several megahertz (MHz) of the radio spectrum. Using Radio-Sky Spectrograph, you will be able to (1) collect, store, retrieve, and edit spectrograph files with header information secured within the file itself, (2) view/show real-time data over the Internet, and (3) [optional] run calibrations from a local calibrator. All software for Radio JOVE is free for noncommercial use.

6.2.2 Antenna Phasing

Another benefit in using a dual dipole array is that you can change the beam direction (beam steering) by adding a phasing cable. By delaying the signal from the south dipole by the proper amount of time, the waves from an emission source collected at the antenna will arrive in phase at the combiner. This extra length of cable delays the signal from the south antenna exactly the same amount as the signal

[1] https://www.radiosky.com/specdownload.html

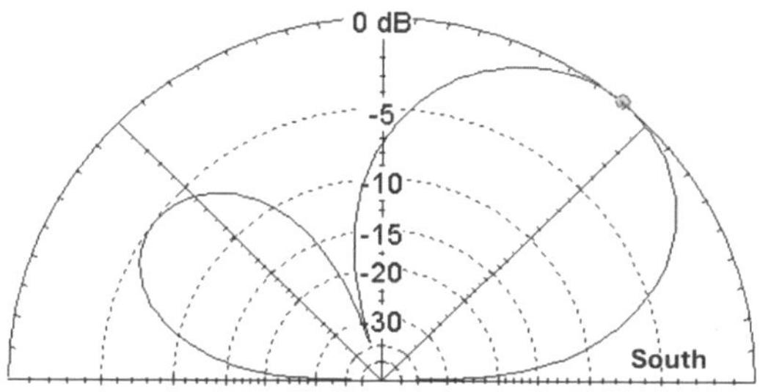

Figure 6.5. A 135° phasing cable added to the south dipole coax steers the antenna beam south. In this example the dipole array is 10 feet high and the beam center is approximately 50° above the southern horizon. Reproduced with permission from Flagg (2005).

to the north antenna was delayed by traveling the extra distance. The extra cable length is called a phasing cable (Figure 6.4), and the delay required in the phasing cable depends on the elevation of the source. If the observer was in the southern hemisphere and the source was in northern declinations, then the beam should be "steered north" by using a phasing cable in the northern antenna (Radio JOVE 2012).

The optimum beam elevation depends upon the latitude of the observer and the declination (or transit elevation) of Jupiter or the Sun. To obtain the strongest signals, antenna phasing and height should be set so that the beam center (maximum gain) is directed toward the transit elevation of the celestial object of interest. For example, Figure 6.5 shows a phasing cable added to steer the beam to 50° above the southern horizon (Radio JOVE 2012).

See Radio JOVE Building Testing:
https://radiojove.gsfc.nasa.gov/radio_telescope/building_testing.php
and (Flagg 2005) for more beaming patterns.

6.2.3 Antenna Effective Area and Beam size

The Radio JOVE single and dual dipole antennas have been modeled using EZNEC Pro+ software (Lewallen 2020). A single dipole with wires running east–west and mounted 10 feet above ground has a north–south half-power beamwidth of about 120° (Radio JOVE 2012). The maximum gain is 5.8 dBi, and compared to an isotropic radiator, $\text{dBi} = 10 \log G$, the gain (G) of the single dipole is 3.8. Therefore, the effective area of a single dipole at 20 MHz is about 68 m^2. For a dual dipole antenna with wires running east–west and mounted 10 feet above ground and no phasing cable, the zenith pointing beam has a half-power beamwidth of about 60° (Radio JOVE 2012). The maximum gain is now 7.8 dBi and the antenna gain $G = 6.0$. Thus, the effective area of the dual dipole antenna is about 107 m^2, more than one-and-a-half times greater than a single dipole. The effective area equation is:

$$A_{\text{eff}} = \frac{c^2 \cdot G}{4\pi \cdot f^2}, \tag{6.1}$$

where G is the antenna gain, c is the speed of light, and f is the frequency in Hertz.

6.2.4 Other Decameter-Wave Radio Antennas

A single or dual dipole is not the only way to observe Jupiter or solar radio emissions. For those requiring a smaller space for an antenna, the two-element Moxon provides modest gain and proven success receiving emissions. A two- or three-element Yagi antenna provides excellent directivity and gain, but usually requires impedance matching and a tower to mount the antenna. Other antennas include a quad, a log-periodic, a helix, and polarimeters. These offer higher performance, but are typically more expensive and require significant mounting structures (Flagg 2005).

As mentioned earlier, Jupiter's DAM radio emissions are highly circularly polarized. Therefore, polarized antennas and antenna arrays are excellent for observing Jupiter and provide more gain than a linearly polarized antenna. A crossed-Yagi or crossed log-periodic antenna gives the ability to detect right-hand (RCP) and left-hand circular (LCP) polarization. A square array of four dipoles can be made into a polarimeter (an instrument that shows RCP and LCP signals separately) by combining the signals from the four dipoles using a hybrid ring.

Finally, receivers with a wide bandwidth can be matched with specialized wide-band antennas providing excellent spectral coverage of Jupiter or solar radio emissions. Wire dipoles allow for 3–4 MHz of bandwidth, but a terminated folded dipole (TFD) can provide 15 MHz of bandwidth, and making a square array can provide excellent spectral coverage and polarization (Typinski 2020). However, adding polarization and beam steering to wide-band systems makes them more complicated and expensive.

6.3 Data Collection, Examples, and Calibration

6.3.1 Data Collection and the Radio JOVE Data Archive

Before making your first observations of the Sun or Jupiter, it is a good idea to set up the antenna and receiver to confirm that everything is working properly. First, of course, follow the complete Radio JOVE 2.1 Setup and Operation Manual (Radio JOVE 2025):

These are some key settings recommended for the best performance. Software, Metadata, and Timing

1. Radio-Spectrograph Software (RSS) version 2.9.77 or newer.
2. SDR Console SDR-Radio v3.3 or newer; SDRuno V1.42 or newer, and SDRc2RSS (SDR console to RSS) software v1.0 or newer.
3. Setting the Metadata in the RSS software is crucial for data comparisons and scientific investigations: Name, Location, Receiver, Antenna, antenna azimuth, and elevation. Be sure to include the height of your antenna, cable length to antenna, phasing cables, computer operating system, weather, local RFI, etc., in the notes section.
4. Set your computer to UT Time by changing your computer clock time zone to Coordinated Universal Time (UTC) and set the time to best possible accuracy given your location – seconds or better. Use a GPS antenna if you do not have internet access.

Data Archive. The Radio JOVE Data Archive[2] has thousands of data files, image files, and sound files of solar, Jupiter, and terrestrial radio emissions for reference. Anyone can search the archive and download data. We encourage and make it easy for you to submit data to the data archive, but a free Data Submitter Account is required to upload data. The maximum file size to upload is 128 MB (about 4 hr of RSS data).

6.3.2 Testing the Receiver and Antenna Together

Before making your first observation, it is a good idea to set up the antenna and receiver to confirm that everything is working properly. For these tests you can set up either a single dipole or the dual dipole array.

1. Basic test: Disconnect and reconnect your antenna.

   Disconnect your antenna – With the software running disconnect your antenna from the SDRplay receiver. The spectrograph will go nearly black (Figure 6.6). You might see some small interference on your chart. Reconnect your antenna – you should see a good color spectrogram like Figure 6.6. Some horizontal bands of interference is normal.

2. Second test — Compare your data with example spectrograms

   Another good test of your system is done by collecting data and comparing it with example spectrograms. Many example spectrograms are

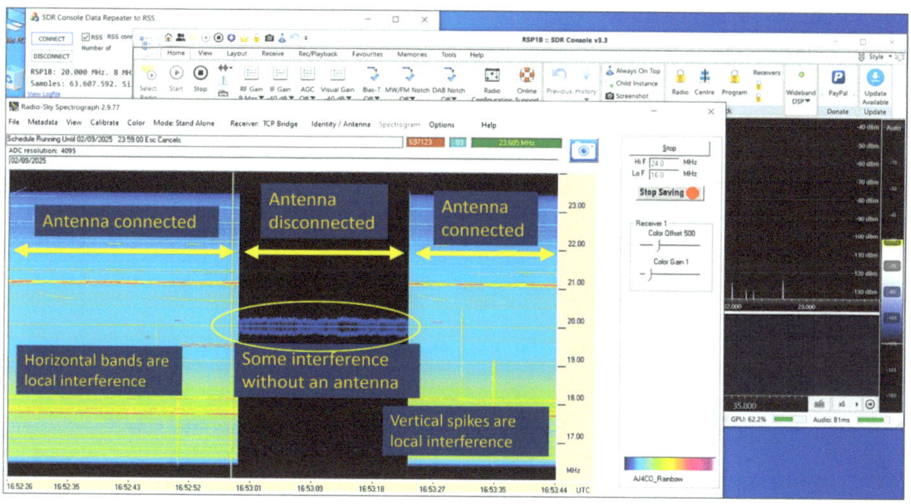

Figure 6.6. Example spectrograph output using RSS display software. Note the SDRc2RSS software window open with the RSS checkbox enabled. The antenna was disconnected and reconnected showing a good color spectrogram.

[2] https://radiojove.net/archive.html

shown here, but many more are found in the data archive or posted daily on the Radio JOVE Groups.io group email listserv.[3]

(a) Radar sweeps – A spectrogram of the ionospheric "sweepers" caused by distant radar used to sound the ionosphere. These appear as diagonal lines on spectrograph graphs from low frequency to high frequency and is a good sign you are picking up distant radio signals. The spectrogram in Figure 6.7 shows 3-4 sweepers and several solar bursts.

(b) The spectrum in Figure 6.8 is a 24 hour graph showing the difference between nighttime and daytime interference. The conditions are fairly quiet at night, but there is more electrical noise during the day, and the ionosphere reflects more radio signals during the daytime. The strong horizontal bands are local broadcast radio stations.

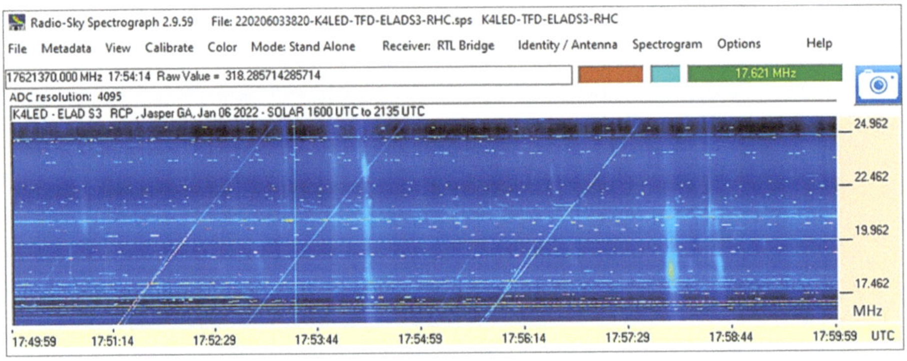

Figure 6.7. An SDR spectrograph from L. Dodd in 2022 showing several solar bursts as the near-vertical "wispy" bands. Several positively sloped "sweeper" radar signals are seen.

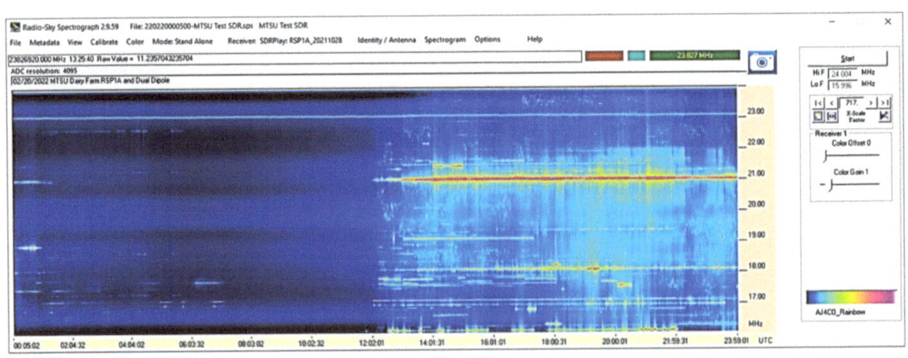

Figure 6.8. Example 24 hour Radio JOVE spectrograph output from C. Higgins in 2022 using RSS display software. Note the clear difference in the nighttime observations on the left and the daytime observations on the right. Local interference is more prevalent during the daytime.

[3] https://groups.io/g/radio-jove

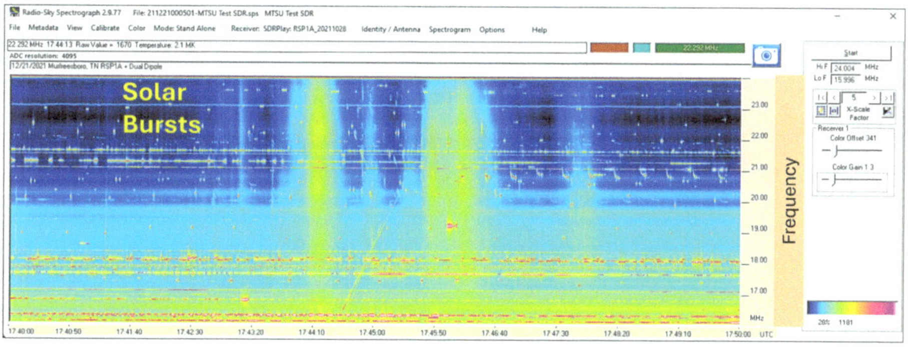

Figure 6.9. A Radio JOVE spectrograph from C. Higgins in 2021 showing multiple solar bursts as the near-vertical "wispy" bands. Spotty interference occurs at several frequencies, with more interference below about 18 MHz. A positively sloped "sweeper" radar signal is seen near the center of the spectrograph.

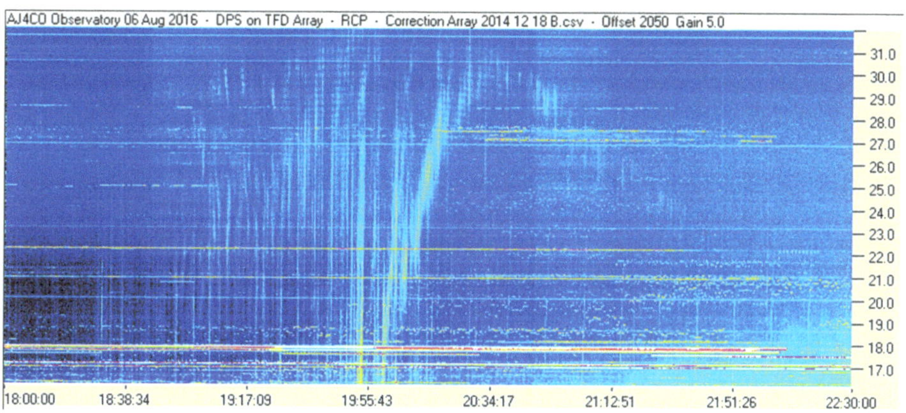

Figure 6.10. A custom built 16–32 MHz wide-band spectrograph shows a Jupiter Io-B emission event over a span of 4.5 hours from D. Typinski in 2016.

(c) Solar bursts show as broadband (near-vertical) "wispy" signals on the spectrograms (Figure 6.9).

(d) Jupiter bursts show as broadband bursts signals sometimes drifting upward or downward on the spectrograms (Figure 6.10). The signals are generally weaker than solar bursts, but usually show more spectral structure.

(e) Terrestrial propagation signals called "tipis" (TPs or teepees) are associated with remote lightning signatures refracting/reflecting off the bottom side of the ionosphere (Figure 6.11).

6.3.3 Calibration

Data generated by a radio telescope must be calibrated to be scientifically useful. To achieve this, the response of the radio telescope to incoming signals must be

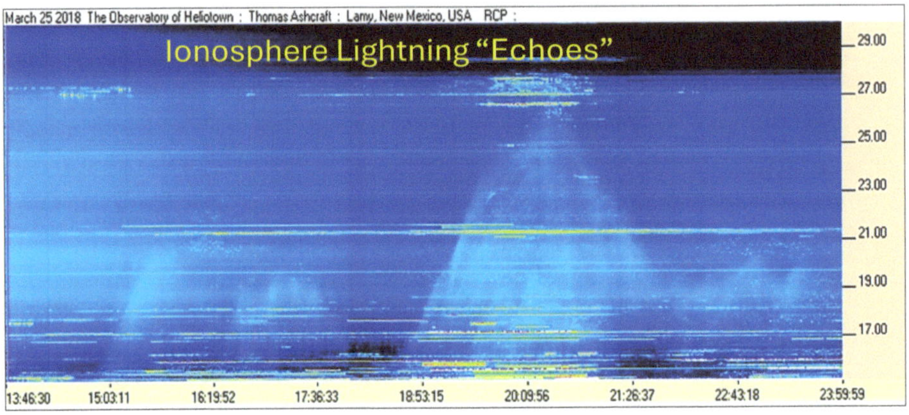

Figure 6.11. A custom built 15–30 MHz wide-band spectrograph from T. Ashcraft in 2018 shows a terrestrial lightning "teepee" signature refracted/reflected from the ionosphere.

characterized and calibrated. This section provides some background material and describes several methods to calibrate data in terms of antenna temperature (T_{ant}).

6.3.3.1 The (Imperfect) Parts of a Radio Telescope

It is useful to look at the various parts of a radio telescope in terms of their response, a measure of its output in relation to its input. In radio electronics, we often desire a linear response; that is, what comes out reflects what goes in, without distortion. With real-world components, however, responses are never perfectly linear.

Three responses are important to a radio telescope: timing, frequency response, and amplitude response. Timing is often provided by a station clock disciplined by GPS. Frequency response describes how the telescope's output amplitude changes as frequency changes when provided with a constant input amplitude. A nice flat frequency response is important for wide-band instruments like spectrographs. Amplitude response describes how the output amplitude of a telescope changes as the input amplitude changes given a constant radio frequency. A distortion-free amplitude response is important for all radio telescope receivers and is of primary concern for single-frequency receivers. We are concerned with amplitude response.

A radio telescope can be thought of as a collection of separate building blocks, most of which influence the response of a telescope (Figure 6.12).

Antennas have responses that vary with frequency. Narrowband antennas can be made to be very efficient, but only at one narrow frequency range — perhaps a few hundred kHz wide. Wide-band antennas used for observation of a wider range of frequencies — say, 15 to 30 MHz wide — have a much flatter frequency response, but still have responses that change somewhat with frequency.

Feed systems consist of everything between the antenna element(s) and the radiometer input. (The radiometer is the instrument that measures the

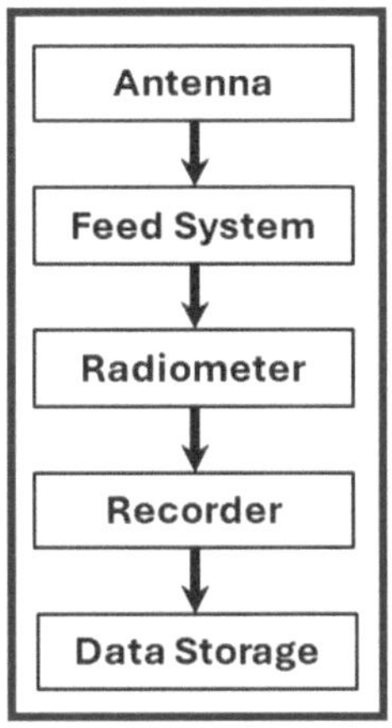

Figure 6.12. A flowchart of the building blocks of a radio telescope.

amplitude of a signal). This includes coaxial cable feed lines, power combiners, preamplifiers, multicouplers, hybrid rings, and switching relays. Coaxial cable attenuates a radio signal to a greater degree as the radio frequency increases. The other components may also have different attenuations or amplifications at different frequencies.

Radiometers are basically a radio receiver without any demodulation function. They not only have a different response at different frequencies, but also have an imperfectly linear response with respect to input amplitude. That is, even at a fixed frequency, the response is close to linear only over a certain range of power levels – this linear response region is the receiver's dynamic range.

Data recorder is generally digital but still contains real-world parts that may contribute to inaccuracies in the response of the radio telescope.

Data storage devices are usually a computer hard drive or solid state memory. This is one component that, when working properly, does not introduce errors into the recorded data.

These imperfect parts work together to measure and record the radio noise amplitude. However, since they are imperfect, the resulting measurements are not perfect. This is why calibration is important. Calibration takes an imperfect response and converts it into something closer to the ideal. A telescope's response will never be perfect, even with strident calibration. However, we require only that the telescope's response be reasonably accurate and the accuracy required depends on the experiment being performed. For Jovian and Solar work in the high frequency (HF) band, an accuracy within 1 or 2 dB is fine. For measuring the cosmic microwave background radiation (CMBR) with the Wilkinson Microwave Anisotropy Probe (WMAP), amplitude accuracy to one part in a hundred thousand was required.

6.3.3.2 Bandwidth Limited Noise – or How a Radio Telescope Works

So far, all cosmic radio signals detected on Earth consist of radio frequency (RF) noise. If we graph radio noise over time, we get something like Figure 6.13(a). Since noise is random, we must somehow be able to quantify this noise.

There is one aspect of cosmic radio noise that is easily quantifiable: its power. The power contained in cosmic radio noise is no different than the power contained in any other radio wave. However, the noise itself is quite different from other radio waves. Most human-generated radio waves are modulated sine waves, and thus are relatively narrow bandwidth. Cosmic noise is most often very wide-bandwidth radio noise — that is, there is no sine wave carrier signal. It is just pure noise.

The power contained in radio noise — called the noise power — may be measured with a device called a total power radiometer or just a radiometer. A radiometer is an instrument not much different from an AC power meter except that it operates at radio frequencies.

All radio telescopes — from Jansky's merry-go-round to the Green Bank Telescope (GBT) and the Very Large Array (VLA) and everything in between, including the Radio JOVE telescope — have at their heart a radiometer. Some operate only at one radio frequency with a narrow bandwidth, while others operate on many frequencies with wide bandwidths concurrently. Regardless of the details, they all do the same thing, they measure and record the voltage present at the feed point of an antenna. This is what radio telescopes do. In fact, this is all that radio telescopes do.

Despite the simplicity of the concept, the engineering can be complex. There is usually some amplification, filtering, and one or more stages of frequency conversion between the antenna feed point and the output of the radio telescope, all in an effort to accurately measure the voltage developed at the antenna feed point.

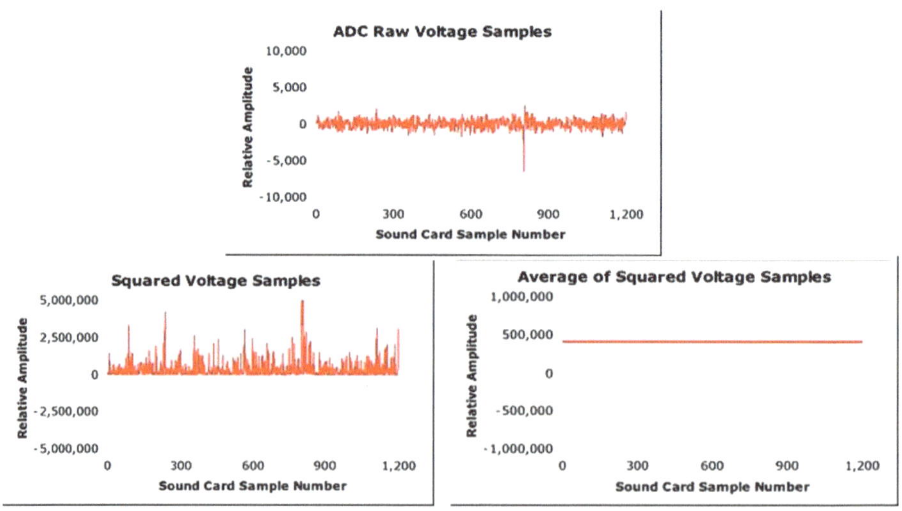

Figure 6.13. (a) A plot of radio noise versus time; (b) the voltages squared from (a); (c) the squared voltages now averaged.

Suppose we have some cosmic radio source — the Sun, Jupiter, Cassiopeia A, the Milky Way galaxy — spewing forth RF noise. This noise impinges on an antenna and in so doing creates a tiny, randomly fluctuating voltage across the antenna feed point terminals. We connect a radio receiver to the antenna feed point to filter and amplify the tiny voltages. We add some method of measuring and recording these voltages, and voilà, we have a radio telescope.

If we are measuring random noise voltages, how come it does not simply average to zero, and how do we get to power? Well, it does average to zero — a look at Figure 6.13 makes that obvious. But that is not how to measure the power in a radio wave — or any other wave, for that matter. For example, if you averaged a bunch of instantaneous voltage readings from the AC mains in your house, the average would be zero since the voltage oscillates between –160 and +160 volts 60 times per second. Obviously, the average voltage says nothing about the power available.

We know from Ohm's law that power is equal to voltage squared divided by resistance; that is:

$$P = \frac{V^2}{R} \tag{6.2}$$

We can measure the voltage many times, very rapidly, then square those readings (Figure 6.13(b)), then average them (Figure 6.13(c)), then divide by R, to obtain power.

Using the AC power as our example again, we know that the voltage oscillates at 60 Hz. If we measure this voltage many times per second (say, 150), then square all the voltage measurements, then take the average, then take the square root, we will get about 115 V. This voltage is known as RMS voltage, for root-mean-square, that is, taking the root of the mean of the squares. We can leave off the square root for power calculations, since the equation for power requires V^2, not V.

But what is R in the power equation? This is something we must calculate. For example, if we know that an incandescent light bulb uses 60 W, then:

$$R = \frac{115 V_{\mathrm{rms}}^2}{60\ \mathrm{W}} = 220\Omega \tag{6.3}$$

Note that this is the resistance of the filament of the light bulb at operating temperature, not cold. We can perform the same calculation for our radio telescope receivers to find R so that we can calculate the power of cosmic radio noise. Instead of a light bulb, we use a radio noise source.

If we have a calibration noise source of known output, and we measure V many times, rapidly, we can solve the power equation for R. Once we know R, we can remove the noise source and replace it with an antenna, then we can calculate the power of the received radio noise. The known output power of the calibration noise source, that is, a noise generator, is measured in units of kelvin – a noise temperature. Noise power is related to noise temperature by

$$P = kTB \tag{6.4}$$

where P is power in watts, T is noise temperature in kelvin, B is the bandwidth in hertz over which the noise is being measured, and k is the Boltzmann constant, which is equal to 1.38×10^{-23} J K^{-1}.

The bandwidth in this equation is the bandwidth over which the receiver measuring the noise power is sensitive. Radio receivers may tune over a wide range, but when tuned to a specific radio frequency, they can only "see" radio waves over a narrow part of the RF spectrum centered on the tuned frequency. For example, a receiver might be tuned to 20 MHz and be sensitive to radio waves with frequencies between 19.997 MHz and 20.003 MHz — in this case we would say that the receiver's bandwidth is 6 kHz.

This is where the term bandwidth-limited-noise, or band-limited-noise, comes from. It is the bandwidth of the receiver that is doing the limiting. If the receiver did not limit the bandwidth, the receiver would be sensitive to RF from DC to daylight, and tuning would be pointless.

Let us try the noise power equation assuming our calibration source has an output noise temperature of 24,000 K (or 24 kK):

$$\begin{aligned}\text{Power} &= kTB = (1.38 \times 10^{-23}\ \text{J K}^{-1}) \cdot (24,000\ \text{K}) \cdot (6\ \text{kHz}) \\ &= 2.0 \times 10^{-15}\ \text{W}.\end{aligned} \tag{6.5}$$

We connect the noise source to the radiometer's input, and, for example, the radiometer measures 710 nV for the RMS voltage for this amount of noise input. (How a radiometer measures this voltage and records it is the topic of the next section.) We can now use the known noise power in watts and the voltage reported by the radiometer to solve for R:

$$R = \frac{V^2}{P} = \frac{(7.1 \times 10^{-7} V)^2}{2.0 \times 10^{-15}\ \text{W}} = 252\quad \Omega. \tag{6.6}$$

This means that we can replace the noise source with an antenna, measure the voltage at the antenna feed point many times, very rapidly, then square those voltages, then average them, then divide by 252 Ω, to obtain power in Watts at the antenna feed point.

Note that the calculated 252 Ω is only a proportionality constant that may not correspond to any real resistance in the physical Universe. Rather, it is the value that is required to make the radiometer's output read correctly for the known input temperature. We used 710 nV as an example. If the same radiometer instead read 430 nV for the same 24 kK input, then R would be something other than 252 Ω.

Since we know how to calculate the power of radio noise, we can simply calibrate the radiometer in terms of temperature directly. Since power and temperature are related by the Boltzmann constant k and bandwidth B, we can easily combine Ohm's law and the noise power equation to derive a more useful relationship for the antenna temperature:

$$T = \frac{V^2}{kBR}. \tag{6.7}$$

For example, if we hook up the 24 kK calibration noise source to the receiver, the radio telescope should measure 710 nV and thus report and record a noise temperature of 24 kK:

$$R = \frac{(7.1 \times 10^{-7}\ \mathrm{V})^2}{[(1.38 \times 10^{-23}\ \mathrm{JK^{-1}})(6\ \mathrm{kHz})(252\Omega)]} = 24\ \mathrm{kK}. \quad (6.8)$$

When the radiometer is connected to the antenna, the calculated noise temperature at the antenna feed point is simply the antenna temperature. It is a noise temperature, often abbreviated as T_{ant}, specifically the noise temperature present at the antenna feed point.

The subject of this section is what to do when the radio telescope invariably reports temperatures other than those it should. That's easy: the telescope must then undergo calibration. In concept, a calibration factor, a_{cal}, is added to the above equation to give us:

$$T = \frac{a_{\mathrm{cal}} \cdot V^2}{kBR}. \quad (6.9)$$

In practice, we ignore k, B, and R altogether and simply adjust a_{cal} so that the radiometer presents the correct reading.

So, after all that, what value is there in measuring the noise temperature accurately? The entire field of radio astronomy — from measuring the variations within the CMBR resulting from density variations present in the early Universe that have since turned into galactic superclusters and great voids, to measuring the relative velocities of neutral hydrogen clouds within galaxies to infer the existence of dark matter, to observing Jovian decametric emission that allows us to deduce how fast Jupiter's core is rotating beneath all its clouds, to the observation of pulsed radio emission proving the existence of rapidly spinning stellar cinders with surface gravity so strong that their protons and electrons are squashed together to form a ball of neutrons as heavy as a star but only 15 km in diameter – all of this is built on nothing more than measuring a bunch of antenna feed point voltages. How cool is that?

6.3.3.3 How to Measure Voltages and Find Power in the Real World

A receiver operating as a total power radiometer (that is, no demodulation) is simply a voltage amplifier, bringing the tiny voltages present at the radio telescope antenna feed point up to a level where they can be measured and recorded. A good receiver will amplify a wide range of voltages without distortion. This distortion-free range is called the dynamic range of the receiver.

Most radio telescopes today measure voltage by sampling the output of the receiver's output noise waveform very rapidly with an analog-to-digital converter (ADC). For example, a simple HF band single channel radio telescope can use the ADCs in the sound card of a computer to measure the output voltage of the receiver.

When using this method, care must be taken such that the sampling rate is at least twice the highest frequency component in the waveform. If we are using a receiver with a 3.5 kHz bandwidth (BW), such as the Radio JOVE version 1 receiver, we would want to take voltage readings at a bare minimum of 7000 times per second. If we do not, we could lose some of the power contained in the higher frequency components of the waveform being measured, and our results would be inaccurate. It is good practice to keep the sample rate to about 2.5 times the receiver BW to allow for the real-world characteristics of the filters in the receiver. The most common sound card sampling rate that meets this requirement for the JOVE receiver's 3.5 kHz BW is 12,000 samples per second (also known as 12 kHz). So, 12 kHz is the sound card sampling rate most often used with the JOVE 1.1 receiver.

Once we have the voltage samples from the ADC, they are simply squared and then multiplied by a calibration factor, the result being a noise power in terms of antenna temperature. We employ a standard noise source to derive the calibration factor.

6.3.3.4 Noise Generators/Calibrators

In order to perform a calibration, one must know the output temperature of the noise generator over the center frequency and bandwidth of the receiver. Common noise generators include the RF-2080 (a fixed temperature of 25 kK), the HP 461 with a step attenuator (1 kK to 75 MK), and the FBX-100 automatic calibrator (1 kK to 100 MK). All the output noise temperatures discussed in this section are at 20 MHz.

Figure 6.14 shows a sample step calibration displayed in Radio-Sky Spectrograph software with an SDR receiver operating over an 8 MHz bandwidth from 16 to 24 MHz. A 12-bit analog-to-digital converter (ADC) gives 4096 colors over a rainbow spectrum. The graph in Figure 6.14 shows the dynamic range of the receiver, that is, the ADC response versus antenna temperature. This system shows a near linear response over 30 dB meaning the receiver is capable of operating from a few kilokelvin up to about 10 MK.

6.3.3.5 Feed System Losses

In order to perform a calibration, one must know all the losses in the system between the sky side of the antenna (the antenna itself can be lossy) and the calibration plane, the point at which the known calibration noise temperatures exist. In our examples, the calibration plane lies at the receiver input connector or very close to it.

Losses may include:

1. Feed line insertion loss (around 2.2 dB per 100 feet for Belden RG-58; around 1.5 dB per 100 feet for Belden RG-59).
2. Feed point to feed line impedance mismatch loss (around 0.1 dB for JOVE dipoles).
3. Balun insertion loss (essentially zero for the JOVE dipoles).
4. Impedance matching transformer insertion loss (not applicable to the JOVE dipoles).

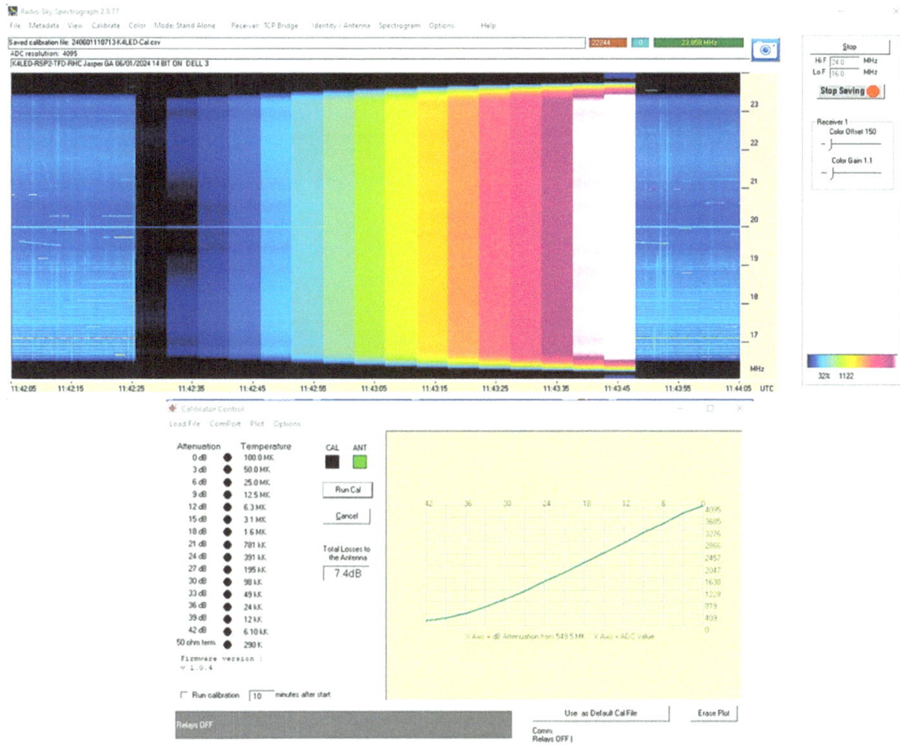

Figure 6.14. A sample step calibration using Radio-Sky Spectrograph software with an SDR receiver operating over a 16–24 MHz bandwidth. The top figure shows the steps associated with a 12-bit analog-to-digital converter (ADC), and the bottom figure shows the dynamic range of the receiver, the ADC response versus antenna temperature. Note the near linear response of the system.

5. Power combiner insertion loss (usually around 0.2 dB per combiner).
6. Feed line to receiver impedance mismatch loss (0.2 dB for JOVE dipoles with 75Ω feed systems, essentially zero for Jove dipoles with 50Ω feed systems).
7. Antenna inefficiency (efficiency is roughly 100% for the dipoles in a JOVE Array).

For example, the losses between the array element feed points and the calibration plane are 3.2 dB. When the calibration noise source is connected near the receiver, the temperature at the calibration plane must be increased by 3.2 dB to find the equivalent antenna temperature. For example, if the calibration noise source has an output temperature of 93 MK, it would take an antenna temperature of 93 MK + 3.2 dB = 194 MK at the feed points to equal 93 MK at the calibration plane. Therefore, when calibrating, one must know not only the noise source temperature but also the losses within the feed system, then one may calibrate in terms of feed point antenna temperature instead of calibration plane temperature.

6.4 Experiments with Radio JOVE

6.4.1 Experiment 1: Observe Solar or Jupiter Radio Bursts

6.4.1.1 Introduction

The Sun can be a very active radio emitter, especially near the time(s) of solar maximum. As mentioned above, solar flares and other magnetic disturbances release plasma into the corona allowing nonthermal emission processes to generate radio waves. At decameter wavelengths, these regions of emission occur about 1–2 solar radii above the photosphere.

Predicting solar bursts is nearly impossible. They can occur many times per day near solar maximum or be zero for many weeks near solar minimum. The best chance for success is to observe many hours per day for many days. If you have the luxury of a semi-permanent setup, you can leave your system running for days or weeks at a time. A single dipole antenna is sufficient to observe solar radio bursts.

Jupiter, on the other hand, is more easily predicted, but more difficult to detect. Jupiter signals are much weaker than solar, so the best configuration is a two-dipole antenna and nighttime observations. The ionosphere is much "thinner" at night, allowing weaker emissions to propagate through. Predicting times when there is a higher probability of radio emissions is easy if you know Jupiter's longitude facing Earth and the orbital position of the moon Io. These Io-related emissions are well known, and software like Radio Jupiter Pro from Radio-Sky Publishing. Radio Sky [4] is an excellent resource to help you find the times to observe. Tables of predictions are also available on the Radio JOVE website.[5]

6.4.1.2 General Procedure

1. Solar
    (a) Set up a single or dual dipole antenna and observe the Sun with your radio telescope for several hours each day.
    (b) If you are using a dual dipole antenna, apply your phasing cable as needed to give you the best results. Figure 6.15(a) from Radio Jupiter Pro shows the green oval beam pattern of a dual dipole antenna and the Sun's location as the yellow icon in the sky over time. In June at mid-northern latitudes, the Sun is high in the sky, and no phasing cable is needed. Jupiter's location is also shown by the brown icon. Six months prior, in January, a phasing cable is required to get the Sun into the antenna beam (Figure 6.15(b)).
    (c) Analyze your data and look for solar radio bursts. Figures 6.7 and 6.9 show examples of solar radio bursts. Good resources to help you identify solar radio bursts:
        i. The best way is to confirm your observations is with data from other observers. Frequent posts are made by Radio JOVE observers on the Radio JOVE Groups.io group email listserv.

[4] https://radiosky.com/rjpro3ishere.html
[5] https://radiojove.net/predictions/

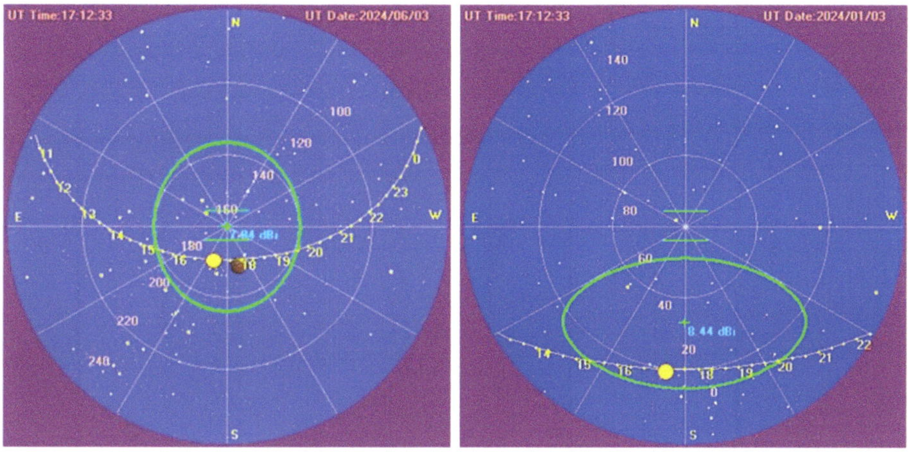

Figure 6.15. Radio Jupiter Pro software plots show a green oval dual dipole antenna pattern on the sky, (a) no phasing so antenna beam is overhead, and (b) phasing cable shifting the beam to the south.

ii. Radio JOVE Solar Radio Emissions Training Module 1.3 or Solar Radio Burst Identification Training Module 3.1: https://radiojove.gsfc.nasa.gov/citizen_science/training.php.

iii. e-Callisto project solar radio burst information: https://reeve.com/Solar/Solar.htm.

2. Jupiter
   (a) Set up a dual dipole to observe Jupiter at night. Daytime observations of Jupiter are possible, but you will have better success at night. Make sure Jupiter is within the beam of your antenna during a predicted Io-related event (Figure 6.16). See Radio Jupiter Pro (https://radiosky.com/rjpro3ishere.html) for help.
   (b) Observe for several hours beginning 30 min before the predicted event.
   (c) Analyze your data and look for Jupiter radio bursts. Figure 6.10 shows an example of Jupiter radio bursts. Good resources to help you identify Jupiter radio bursts:
      i. Confirm your observations with data from other observers on the Radio JOVE Groups.io group email listserv.
      ii. See Radio JOVE Jupiter Radio Emissions Training Module 1.4: https://radiojove.gsfc.nasa.gov/citizen_science/training/TM1.4_Jupiter_Radio_Emissions.pdf.

6.4.1.3 Questions

1. Explain why Jupiter is not seen in the sky in Figure 6.15(b).
2. What type of solar bursts are shown in Figure 6.9?
3. Identify the types of solar bursts seen in Figure 6.17.

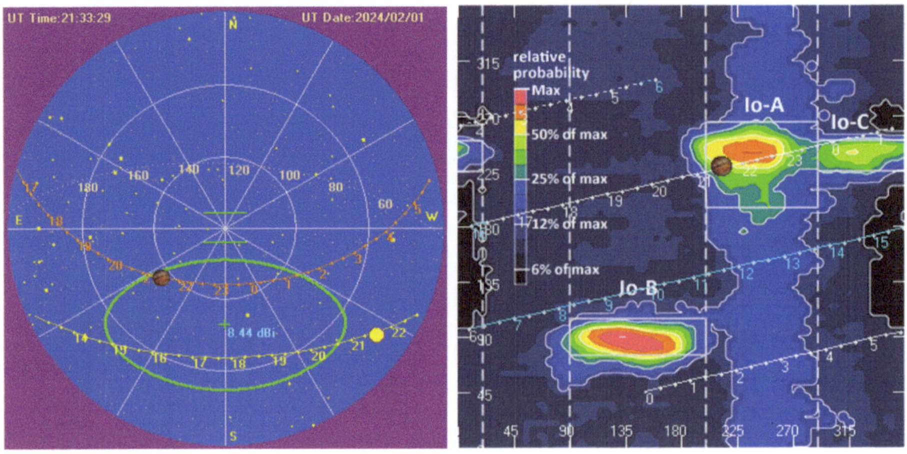

Figure 6.16. Radio Jupiter Pro software plots, (a) Jupiter entering the antenna beam, (b) Jupiter radio prediction Io orbital phase versus Jupiter longitude map showing Jupiter entering the high probability Io-A zone at 2100 UTC.

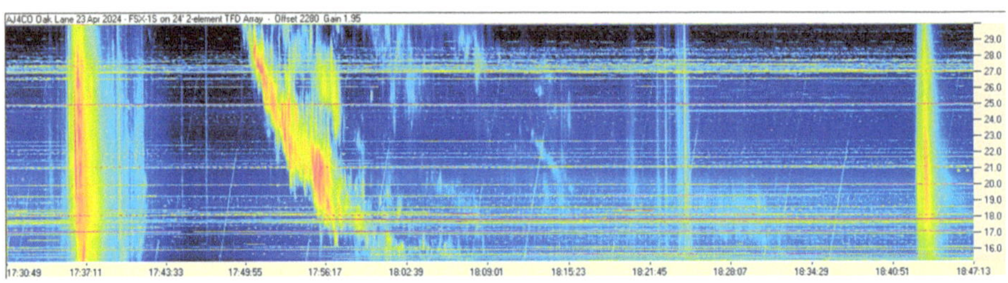

Figure 6.17. Solar radio bursts on 2024 April 23 from D. Typinski.

6.4.2 Experiment 2: Observe Terrestrial Lightning Propagation “Tipi” Signatures

6.4.2.1 Introduction

A high frequency spectral feature has been identified in ground-based spectrographs and recorded by a group of citizen scientists from the Radio JOVE project (Fung et al. 2020). This feature is a tipi (TP) shape found in data between 15 and 30 MHz, where the spectral enhancement frequency increases and then decreases with time (Figures 6.18 and 6.11). The presence of these features is currently being attributed to ionospheric reflection of radio emissions from lightning activities in remote thunderstorms moving relative to the observer.

6.4.2.2 General Procedure

1. Observe over several days/weeks to observe a tipi (TP) event.

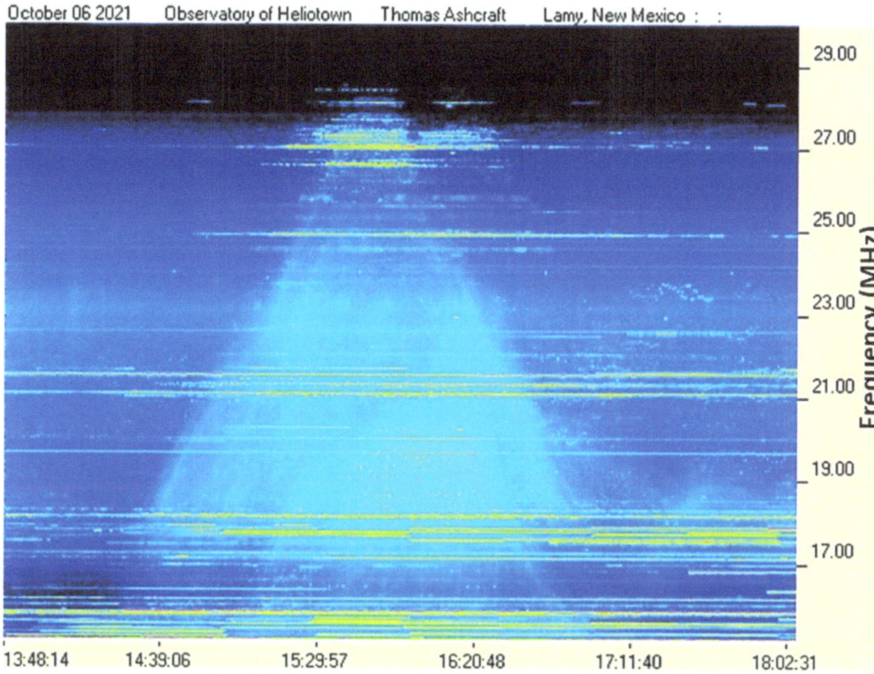

Figure 6.18. A propagation TP on 2021 October 6, from T. Ashcraft, Lamy, NM.

2. Generally, these events occur later in the afternoon when distant thunderstorms occur, but they can occur any time during the day. They rarely occur at night.

6.4.2.3 Questions

1. What is the duration of the TP event? The answer depends on the frequency, and sometimes the lower frequency is difficult to determine due to interference.
2. What is the maximum frequency?
3. What are the slopes of the leading and trailing edges of the TP event? The slopes can give insight into the motion of the thunderstorm; see Fung et al. (2020) for discussion and analysis.
4. Can you determine a possible thunderstorm that might have caused this event? Good resources for lightning maps:
    1. https://map.blitzortung.org
    2. https://www.lightningmaps.org

6.4.3 Experiment 3: Measure the Galactic Background Temperature

Part of this section has been reproduced with permission from Typinski (2017).

6.4.3.1 Introduction (This Experiment Requires a Calibrator)

The purpose of this experiment is to determine the background noise level at your observing site and thus how radio quiet it is. Therefore, you will know if your receiving site is suitable for monitoring radio emissions from Jupiter, the Sun, and the Galactic background. This experiment requires a calibrator (see Calibration above). You can build your own calibrator; the complete instructions are here:

https://radiosky.com/jovecal/JOVE_Cal1_Manual.pdf

There are many sources of interference that can completely mask the cosmic signals you are trying to detect. Radio Frequency Interference (RFI) sources include arcing power line insulators, electrical fences, electrical machinery, computers, fluorescent lights, solar panel inverters, grow lamps, etc. Locate the antenna as far away as possible from noise sources. A 96 foot length of RG59 coaxial cable (three wavelengths) is provided to run from the antenna to the location of JOVE receiver, computer, and calibrator. If you need a longer cable run, the total cable length should be multiple half-wavelengths long. See the JOVE antenna manual for additional guidance:

https://radiojove.gsfc.nasa.gov/radio_telescope/manuals/RJ_Antenna_Manual_v2.1_2025_02_07.pdf

At a perfectly radio-quiet site, the background noise level will be set by the Galactic background temperature. At 20 MHz this is about 50 kK plus or minus a couple of dB depending on whether the Galactic plane is in the antenna beam. If your observing site is radio quiet, your nighttime background temperature should run in the 25–75 kK range. Locally generated radio frequency noise and even distant stations and other forms of interference will add to the galactic background temperature. Daytime background temperatures are usually much higher than nighttime temperatures due to distant noise sources and terrestrial stations whose signals propagate in from afar. If the background temperature is higher than 200 kK, you will only be able to detect the strongest Jupiter storms, and if it exceeds 500 kK, you may only be able to detect the strongest solar bursts.

6.4.3.2 General Procedure

As explained above, a noise calibrator allows us to rescale the raw voltage units so that the output is given in absolute units of antenna temperature (Figure 6.19).

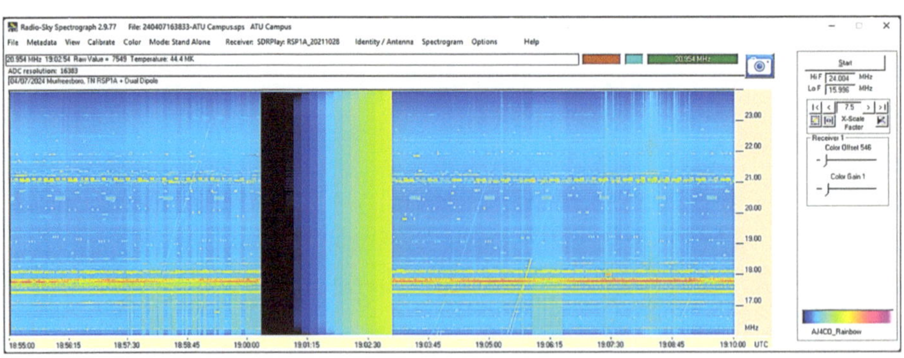

Figure 6.19. Example Radio-Sky Spectrograph showing a step calibration. The raw data values are converted to Antenna Temperature in units of kilokelvin (kK).

Measuring the Background Level. Observe for several hours after sunset and note the background temperature.

6.4.3.3 Questions

1. See Figure 6.14. The calibration steps are 290 K, and 6, 12, 24, 49, 98, 195, 391, 781, 1600, 3100, 6300 kK, and 12.5, 25, 50, and 100 MK. Based on the color of the background before/after the calibration, estimate the Galactic background temperature.
2. Repeat for the data in Figure 6.19.
3. Estimate the temperature of the strongest solar bursts in Figure 6.17.

6.4.4 Experiment 4: Estimate the Total Power Emitted from a Solar Radio Burst

6.4.4.1 Introduction

The Sun is the brightest radio source in the sky, and when a major flare occurs, it can release a tremendous amount of energy in all parts of the spectrum. A solar flare typically results in one or many Type III solar radio bursts (Figures 6.7, 6.9, and 6.17). By measuring a calibrated peak amplitude of a Type III solar radio burst, one can estimate the total power emitted from the burst.

6.4.4.2 General Procedure

1. Observe the Sun over several days/weeks to observe a solar radio burst.
   (a) Alternate — use the Radio Jove data archive: https://radiojove.net/archive.html
   (b) Alternate — find data from Jim Brown on 2019 May 01 and look at the different solar burst files – On the right there is an image file, a Skypipe file (.spd), and maybe a .txt file. Choose the Radio-Skypipe data file.
2. Copy the file to your computer and load the .spd file on your Radio-Skypipe program. The Filename = 1556842475UT190501172527_HNRAO_1. Ignore the first set of numbers, but see the date and time marked as UTyymmddhhmmss_observatory name (2019 May 01, 17h 25m 27s and Hawk's Nest Radio Observatory in Industry, Pennsylvania)
3. Use the analysis buttons within Radio-Skypipe to stretch the scale. Carefully measure the peak intensity (in kelvins) and the total duration of the largest burst. Also measure the peak and duration of the smallest burst (see 17:28 UT).
4. For each solar radio burst, compute the following:
   (a) The intensity for a blackbody at this peak temperature.
   (b) Flux density (treat the source size as the same size as the Earth).
   (c) Total Power in Watts.
   (d) Total energy released.

6.4.4.3 Questions

1. What assumptions are you making about the isotropic nature of the emission?
2. In general, why do you think that stars are not strong thermal emitters of radio emission?

6.4.5 Experiment 5: Determine a Quiet Day Curve (QDC) of the Galactic Radio Background

6.4.5.1 Introduction

The goal of this observational project is to determine the long-term average diurnal radio background level at your location, adapted from Fung (2020).

A quiet-day curve (QDC) is a long-term measurement of the galactic radio background noise after removing short-term and local interference. In doing so, one can successfully identify, isolate, and to the extent possible minimize local RFIs, and finally obtain an extended observation interval a net 24 hour radio background level curve for your station location.

The naturally occurring cosmic and galactic radio noise background is due to the combination of radio signals from all the cosmic and galactic radio sources. It permeates the radio sky so it cannot be avoided. Although the radio sources within our own Milky Way Galaxy are intrinsically weaker than some of the cosmic radio sources like quasi-stellar objects (QSOs) and radio galaxies, the galactic radio sources are much closer to us, so they appear stronger. Consequently, the natural radio background that we must deal with is primarily due to the galactic radio sources concentrated in the Galactic plane and near the Galactic Center.

Since there is no avoiding the Galactic radio background when we make radio observations of Jupiter and the Sun, the signals received by our telescopes at any time are always the sum of the background signals and the signals from our target, Jupiter, or the Sun. Fortunately, the emissions from either the Sun or Jupiter are generally stronger (perhaps due to their proximity) than the background signals, and they are easily distinguishable. To be able to ascertain the intrinsic strengths of the observed emissions (so that we can gain a better understanding of the physical processes at the Sun and Jupiter), we need to remove the contamination of the background radiation levels from our raw data.

Figure 6.20 shows an example of raw Radio JOVE observations at 20.1 MHz over a 24 hr interval on 2020 April 10. The horizontal axis provides the observation times in UTC, and 0500–1000 UTC corresponds to 0–5 AM when both the local RFI and the Sun should be relatively quiet, such that the remaining signals are primarily due to the galactic radio background. The interval labeled "Galactic Hump" indicates the times of transit of the Galactic Center over the observing location or through the antenna beam. The strong signals above the background during daytime hours are caused by significantly higher levels of RFI and solar radio emissions detected during the day.

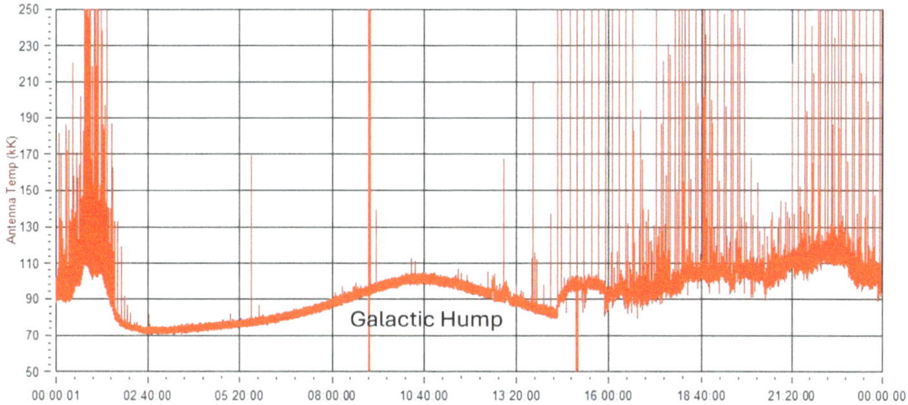

Figure 6.20. An example 24 hour plot of 20.1 MHz radio observations on 2020 April 10 from D. Typinski. With no other radio sources or interference, the background is the lower limit is the Galactic radio background. At about 1100 UTC, the center of the Galaxy drifts through the antenna beam forming a curved "hump."

In order to obtain the full 24 hour quiet-day curve of the galactic radio background, an extended interval of nighttime observations is required, so that the entire quiet-day curve could be obtained under low-noise conditions.

6.4.5.2 General Procedure

1. Make a 24 hour observation producing a plot like Figure 6.20. A single channel of a radio spectrograph can be plotted easily in Radio-Skypipe software:
   https://radiosky.com/skypipeishere.html
   or using Autoplot software:
   https://autoplot.org/
2. Record the system configuration, usually saved in the metadata for each file.
3. Repeat the 24 hour observing run for several weeks, ideally for 6–12 months.
4. If possible, use a calibrator to obtain the best calibration of the instrument to measure signal strengths in antenna temperature or other physically meaningful units.
5. Data analysis can be found in "The Quiet Day Curve" (Typinski 2017).

6.4.5.3 Questions

1. Why does the peak of the Galactic hump move 4 minutes earlier each successive day?
2. Data analysis requires data to be recast in sidereal time. Using Question 1 as a hint, why do you think this is important?
3. In what range does the galactic background temperature fluctuate in Figure 6.20?

References

Flagg, R. S. 2005, Listening to Jupiter (2nd ed.; Cambridge, MA: Radio-Sky Publishing)

Fung, S. 2020, Determining the Quiet-Day Curve of the Galactic Radio Background radiojove.gsfc.nasa.gov/citizen_science/projects/2020/Citizen_Science_Project_1.0.pdf, Radio JOVE

Fung, S., Typinski, D., Flagg, R., et al. 2020, GeoRL, 47, e2020GL087307

Gopalswamy, N., Mäkelä, P., Yashiro, S., et al. 2018, ApL, 868, L19

Radio JOVE 2012, The Radio JOVE Antenna Manual Website radiojove.gsfc.nasa.gov/radio_-telescope/building_testing.php

Radio JOVE 2025, Radio JOVE 2.1 Setup and Operation Manual Website radiojove.gsfc.nasa.gov/radio_telescope/building_testing.php

Lewallen, R. 2020, EZNEC Pro+ v7.0 Antenna Software by W7EL www.eznec.com

Typinski, D. 2017, The Quiet Day Curve SARA Journal

Typinski, D. 2020, TFD Square Array AN-TFD-24-4 Instruction Manual radiojove.net/SUG/

Part III

Large Facilities/Missions That Include Student Participation in Instrumentation

Chapter 7

The Long Wavelength Array

Sarah I. Chastain, Gregory B. Taylor and Jayce Dowell

Although the Long-Wavelength Array (LWA) is a large-scale, open-skies radio telescope with calls for proposals, etc., its history of student involvement in its construction and operation makes it an ideal project for radio astronomy data analysis in the classroom. Here we present several exercises with publicly available LWA data that explore a variety of astrophysical and space weather phenomena, some of which are difficult to observe with small-scale radio telescopes: pulsars, Jupiter bursts, solar radio bursts, and meteor radio afterglow (MRA) trails. *This chapter includes pure data analysis exercises not requiring equipment.*[1]

7.1 Introduction

The LWA is a low-frequency radio interferometer operating between 3 and 88 MHz (Taylor et al. 2012). It consists of multiple independent stations that can be combined together through interferometry to make what is called the "LWA Swarm," which is the largest radio interferometer in this frequency range in North America. As of this writing it consists of three stations currently operating, two of them with 256 dipoles located in New Mexico (located at Sevilleta National Wildlife Refuge and the center of the Very Large Array, Figure 7.1) and one in Owens Valley, California that has recently been upgraded to 352 dipoles. An additional mini-station (Taylor et al. 2025) with 64 dipoles is constructed and has undergone commissioning activities at the end of the north arm of the Very Large Array (VLA). Two more mini-stations, one in Comanche Springs, Texas and one located in Arizona, are currently in the planning stages and expected to start construction soon. The dipoles utilized are LWA antennas, originally developed by the Naval Research Laboratory (Hicks et al. 2012). They are used in several other of the radio observatory designs that are shown in this book: the Deployable

[1] The original tutorials, which were written for internal usage on the LWA's computing cluster, can be found at https://lda10g.alliance.unm.edu/tutorial/.

doi:10.1088/2514-3433/ae1223ch7

Figure 7.1. The first LWA station located at the NRAO Very Large Array in New Mexico. Credit: Long Wavelength Array Collaboration

Low-Band Ionosphere and Transient Experiment (DLITE; Chapter 10), the SunRISE Radio Ground Lab (Chapter 9), and the Low-Frequency All-Sky Monitor (LoFASM; Chapter 11).

7.2 Hardware Requirements

The projects in this chapter require a computer with a large amount of storage. The storage requirements are as follows:

- Pulsars at Low Frequencies: 27 GB.
- Studying the Transient Universe: 303 GB.
- All-Sky Imaging: 5 GB.

The processing itself should not be too resource intensive and most modern computers should suffice.

7.2.1 LWA Data Products

The Long Wavelength Array has different observing modes that it operates in. Some of these modes are specific to a particular station design and others may be deprecated by newer modes in the future. Each mode is designed to deliver a data product for certain kinds of science cases. There are various modes that can be used depending on the science goals. Currently the modes that are available vary slightly depending on the station. The Transient Buffer Wideband (TBW) is a mode specific to the first LWA station that records the full bandwidth for all dipoles for 61 ms every 5 min. The Transient Buffer Narrowband (TBN) is a mode common to all stations that provides 75 kHz of bandwidth from all dipoles continuously. The Transient Buffer frequency-domain is a mode present on the second LWA station and future stations that is similar to the TBW in that it provides 20 MHz of bandwidth in 1–2 s dumps every 3 min. All stations can form beams (DRX mode) with up to 20 MHz bandwidth and two tunings. Details of the number of beams and flexibility of the tunings vary according to station. Finally, the LWA Swarm mode uses all the stations in beam-forming mode and then correlates the beams to synthesise a large aperture for higher resolution imaging.

7.3 LWA Tutorials

These tutorials are intended to demonstrate a selection of the kinds of science that can be done with the LWA and how to observe with the LWA. It is by no means comprehensive; notably omitted is a demonstration of lightning observations or of LWA Swarm observations. What is present is a variety of tutorials involving the kinds of data products that can be generated from a single LWA station and how to generate an observation session.

These tutorials depend on the following software:

- Apptainer, formerly known as singularity: https://apptainer.org/.
- The tutorial data found at: https://lda10g.alliance.unm.edu/tutorial.
- The tutorial container found at: https://lda10g.alliance.unm.edu/tutorial/lwa_tutorial.sif.

To start the container run:

```
apptainer shell lwa_tutorial.sif
```

Then navigate to the folder where you have placed the tutorial data. If you are using Windows, you will need to run this within Windows Subsystem for Linux, which is also required for installing apptainer.

7.3.1 Observing with the LWA

Space required: None

1. Start the LWA observation scheduling GUI that is part of the SessionSchedules extension with:

   ```
   sessionGUI.py -lwasv
   ```

   This will bring up a window that will help in preparing the session definition files (SDFs) used for scheduling observations at LWA-SV. Running this command without `-lwasv` will bring up the settings for generating an observation at LWA1.
2. Define a new set of observations using the `File -> New` menu or by clicking the "New Project" icon. This will open up the observer and project information window shown in Figure 7.2. In this window you can define your name, the project, the LWA-SV mode to be used to collect the data, whether or not to use the data recorder spectrometer, and how the data are to be returned to you. The nomenclature is that a project is made of one or more sessions. Each session uses one of the LWA-SV outputs (two DRX beams and TBF or TBN) and consists of multiple sequential observations. Note that TBN may be deprecated in the near future. For DRX observations

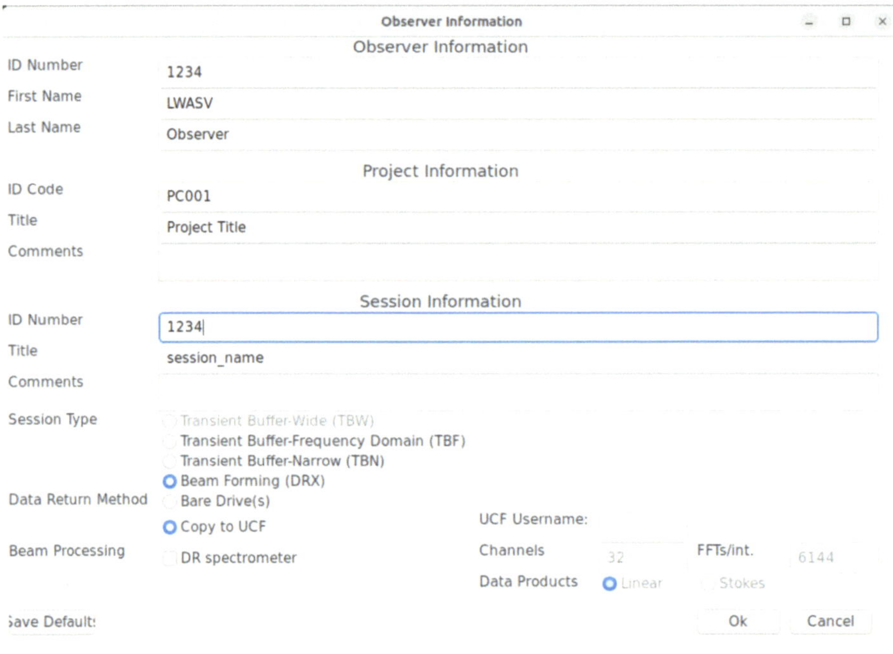

Figure 7.2. The sessionGUI.py observer and project information window. Here you can define your name, the project, the LWA-SV mode to be used to collect the data, whether or not to use the data recorder spectrometer, and how the data are be to returned to you. Credit: Long Wavelength Array Collaboration

you can choose to observe a static position or you may track an RA/Dec. position, the Sun, or Jupiter. It is a good idea to check the current list of observing constraints posted[2] to make sure that there are no issues that would interfere with the observations. For the observations of B0329+54 you will want to select the DR spectrometer mode. For the purpose of this tutorial you should select "Bare Drives(s)," however for a real observation you will likely want to choose "Copy to UCF" instead. The default values for the number of channels is too small and the integration time, the number of FFT windows per integration, is too long to resolve the pulse. You should specify 1024 channels of this observation and reduce the number of FFT windows per integration down to a value that is equivalent to about 40 ms. The currently supported modes for the data recorder spectrometer are also available.[3]

3. After defining the observer information, create an observation to look at the pulsar B0329+54. This can be done through the `Observations -> Add -> DRX - RA/Dec` menu or the "New RA/Dec Target" icon **R**. This will add a new blank observation to the window. To find the coordinates of

[2] http://www.phys.unm.edu/lwa/astro/currentissues.html
[3] http://www.phys.unm.edu/lwa/astro/scheds/spec.html

B0329+54, enter the source "PSR B0329+54" into the target field, select the observation via the check box on the left side of the screen, and choose `Observations -> Resolve Selected`.

4. We will now check the visibility of the source. First, change the duration to 24 hours. Then select `Observations -> Session at a Glance`. This will plot the elevation of the source throughout the observation.
5. Define the desired parameters of the observation (observing time, tuning, bandwidth, etc.). For the DRX mode each pointing has two independent tunings that can be tuned anywhere in the 10–88 MHz band of LWA-SV. To get the list of filter codes, use the `Help -> Filter Codes` dialog. For pulsar observations it is good to space the tunings by the bandwidth so that the two tunings can be combined together and jointly analyzed, e.g., if you are using filter #7, space the tunings by 19.6 MHz.
6. Check the data volume of your observations using the `Data -> Estimated Data` Volume feature. This will display a window with the raw data volume for each of the observations as well as the session total.
7. Once you are happy with the observation setup, save the observation to disk. Before the file is saved it will be validated. If any of the observations have invalid parameters, they will be highlighted in red. The terminal window will also list the invalid parameters. Be sure to correct any problems before you exit `sessionGUI.py`.
8. SDFs are submitted to LWA station for observation through the SDF validator.[4] This page accepts multiple SDFs or a tarball containing multiple SDFs and validates each file. Any errors in the SDFs will also be noted as part of the validation process. Valid files are sent to the observing queue to be run by the LWA station operator.

7.3.2 Pulsars at Low Frequencies

7.3.2.1 Observing a Known Pulsar in Spectrometer Mode

Space required: 918 MB

1. Take a look at the power and clipping levels in the spectrometer data file using scripts in the Commissioning repository. You can get an overview of the power levels and clipping via:

   ```
   drspecFileCheck.py 056770_000044687
   ```

   which will report both quantities. Optimal values of power are between seven and ten. Below seven some dynamic range is lost because the signal strength is not enough to populate many bit levels in the data. Above ten the clipping starts to become severe and the dynamic range is also compromised. However, high values for the power and clipping may be transient, e.g., radio frequency interference (RFI), and do not necessarily mean that the observation is useless. If the power or clipping is high for either tuning,

[4] http://fornax.phys.unm.edu/lwa/validator/index.html

convert the data to an HDF5 file and interactively plot the results using:

```
drspec2hdf.py 056770_000044687
plotHDF.py 056770_000044687-waterfall.hdf5
```

Inside plotHDF.py you can view both tunings and polarizations in a variety of ways and look at how the data evolve in time and frequency.

2. Now convert the file into PSRFITS format in preparation for folding the pulsar using PRESTO.

```
writePsrfits2FromDRSpec.py -s B0329+54 056770_000044687
```

This will create two files:

```
drx_56770_B0329+54_b2t1_0001.fits
drx_56770_B0329+54_b2t2_0001.fits
```

One for each tuning. If the tunings are spaced by the sample rate the two PSRFITS files can be combined with combine lwa and jointly analyzed in PRESTO:

```
combine_lwa2 -o drx_56770_B0329+54_b2 \
drx_56770_B0329+54_b2t2_0001.fits \
drx_56770_B0329+54_b2t1_0001.fits
```

This will create a new file named "`drx_56770_B0329+54_b2_0001.fits`" that contains the data from the two input files.

3. Although the LWA1 site is relatively free of RFI it is often a good idea to create a mask to remove any RFI that may be present in the data. To create this mask use the PRESTO rfifind utility:

```
rfifind -time 2.0 -o \
mask_B0329+54_b2 drx_56770_B0329+54_b2_0001.fits
```

The resulting mask includes a postscript file that can be viewed with evince or ghostscript.

4. Finally, incoherently dedisperse the PSRFITS file and fold the data at the pulsar's period of ≈715 ms using the PRESTO prepfold utility:

```
prepfold -psr 0329+54 -mask mask_B0329+54_b2_rfifind.mask \
-nsub 128 drx_56770_B0329+54_b2_0001.fits
```

After folding, prepfold will generate an analysis plot showing the pulse strength over time, the constraints on the period and period rate of change, and the constraints on the dispersion measure (DM). Press the "enter" key in the terminal window when finished.

7.3.2.2 Observing an Unknown Pulsar in Raw Data Mode

Space required: 18 GB

1. Start by converting the raw DRX data file to PSRFITS via:

```
writePsrfits2.py 056227_000024985_DRX.dat
```

   and then create RFI masks for the two PSRFITS files similar to before. For this example we will name our two rfi masks "mask1" and "mask2."

2. Since the DM of the pulsar is only constrained between 5 and 15 pc cm^2, search both files with PRESTO by first incoherently dedispersing files at a collection of trial DMs.

```
for i in {1..2}
do
 prepsubband -lodm 5 -dmstep 0.1 -numdms 100 \
 -nsub 128 -mask mask$i_rfifind.mask -numout ##### \
 -o band$i drx_56227_None_b4t$i_0001.fits
done
```

   This script will create a collection of dedispersed files with a base name of "`band$i`" where `$i` is either 1 or 2. You can figure out the value used for `-numout` by looking at the number of points that rfifind reports when it starts, and then determining the closest power of 2. The next largest power of two can be found by running the following lines in Python:

```
import math
print(2**math.ceil(math.log(######/math.log(2))))
```

3. For each trial DM file search for a pulsar using:

```
accelsearch -zmax 0 -numharm 16 <name_to_search>
```

   Since prepsubband generates many files it is convenient to process them all at once with the xargs command:

```
ls <output_base_name>*.dat | xargs -n1 accelsearch \
-zmax 0 -numharm 16
```

   which will run accelsearch on each file found by the ls command. Sift through the various pulsar candidates found by accelsearch to find the approximate DM and period for the pulsar using the PRESTO utility ACCEL_sift.py:

```
python $PRESTO/examplescripts/ACCEL_sift.py *.cand
```

   The pulsar should show up across several closely spaced DMs at the same period.

4. Once the approximate DM and period are known, fold the original data at the pulsar's period. This searches over a small range of DM and period to get the best fit. Note: prep fold takes DM values in pc cm^{-3} and periods in seconds. Note that you may need to convert the period from milliseconds into seconds.

```
prepfold -dm ##.# -p ##.# psrfitsfilename -nsub 128 \
-mask $maskname
```

Where you would replace `$maskname` with the appropriate maskname from above, such as mask1.mask. Note the final DM and period.

7.3.2.3 Improving the Detection Using Coherent Dedispersion

Space required: 5.8 GB

1. Up to this point only incoherent dedispersion has been used for the pulsar analysis. However, the dispersion introduced by the interstellar medium can be removed from the raw data more precisely using coherent dedispersion if the DM is known ahead of time. To apply coherent dedispersion to LWA1 DRX data use the writePsrfits2D.py script:

```
writePsrfits2D.py ##.# 056227_000024985_DRX.dat
```

where the first argument to the script is the DM found above. After the file has been converted, create RFI masks for the two PSRFITS files similar to Sections 3.1 and 3.2.

2. Fold the data at the pulsar's period using prepfold. This will also search over a small range of DM and period to get the best fit.

```
prepfold -dm ##.# -p ##.# psrfitsfilename -nsub 128 \
-mask $maskname
```

Note the final DM and period. Is the detection stronger with coherent dedispersion?

7.3.2.4 Measuring Faraday Rotation

Space required: 1.4 GB

1. The data for this exercise originally came from the LWA Pulsar Archive.[5] This archive contains observations of the 107 pulsars detected using LWA1 and has data that spans several years for some pulsar. The data on PSR B0950+08 comes from January of 2018 and is at a frequency of 64.5 MHz. All of the additional data available for this pulsar on this date can be found at http://lda10g.alliance.unm.edu/PulsarArchive/B0950+08/58132/.
2. Since many of the pulsar analysis tools used in this exercise default to writing in the same directory as the data it is convenient to create a local copy of

[5] http://lda10g.alliance.unm.edu/PulsarArchive/

data. Do this by copying all the contents of the https://lda10g.alliance.unm.edu/tutorial/PulsarRM/ directory. Now make median_6.psh and plot_pol executable:

```
chmod +x median_6.psh
chmod +x plot_pol
```

3. Using your local copy of the data, flag obvious radio frequency interference (RFI) using the median 6.psh script.

```
./median_6.psh 58132_B0950+08_64.5MHz.0000.ar
```

This script finds and flags data that are more than 6σ from the median and writes the results to a new file with a .zz extension
4. Next, integrate over all the periods (time-bins) of the pulsar. This can be done using the PSRCHIVE command pam:

```
pam -T -e tscrunch 58132_B0950+08_64.5MHz.0000.zz
```

Here, the `-T` flag tells pam to integrate over times and the `-e` flag sets the output extension to `.tscrunch`.
5. In order to convert to Stokes parameters we will again make use of the pam command, this time with the `-S` flag:

```
pam -S -e tscrunchS 58132_B0950+08_64.5MHz.0000.tscrunch
```

The four Stokes parameters (I, Q, U, and V) can be plotted to view the effects of Faraday Rotation using:

```
./plot_pol 58132_B0950+08_64.5MHz.0000.tscrunchS
```

The Faraday Rotation manifests as frequency-dependent banding in the Stokes Q (second plot from left) and U (third plot from left) plots of frequency versus pulsar phase.
6. Now attempt to fit a rotation measure (RM) to the data. We know the RM of PSR B0950+08 is $\approx$2, but the ionosphere can change that up to $\pm$1.5, so we will look at RMs from –5 to 5 and test 256 values of RM to get a good fit:

```
rmfit -D -K 58132_B0950+08_64.5MHz.0000_rmfit.eps/CPS \
-m -5,5,256 58132_B0950+08_64.5MHz.0000.tscrunch
```

This will create a plot of trial RMs versus total polarization and attempt to fit a Gaussian to it. The plot is stored as a .eps file that can be viewed with evince.
7. Using the best-fit value for the RM, apply it to the data and re-generate the plot of the Stokes parameters:

```
pam -R #.### -e tscrunchRM 58132_B0950+08_64.5MHz.0000.
tscrunch
pam -S -e tscrunchRMS 58132_B0950+08_64.5MHz.0000.tscrunchRM
./plot_pol 58132_B0950+08_64.5MHz.0000.tscrunchRMS
```

In the first command of the above "#.###" is the best-fit RM value found with rmfit.

7.3.3 Studying the Transient Universe

7.3.3.1 Jovian Bursts

Space required: 273 GB depending on selection in step 2. For step 1 only: 91 MB

1. The data being analyzed here are part of a commissioning test observation of Jupiter. Unlike most LWA1 beam formed data, this file contains a single 19.6 MS s^{-1} tuning. Since reducing the raw data, even at a reduced spectral and temporal resolution, takes about two hours, a pre-reduced file named

```
055917_000007389_Jupiter-waterfall-low.hdf5
plotHDF.py 055917_000007389_Jupiter-waterfall-low.hdf5
```

   and look for evidence of bursts during the file. For each burst note the time offset, UTC start time, and duration. The plotting script also allows for the color map, stretch, and scale to be interactively adjusted which may help identify events.
2. After you have identified any bursts in the file, produce higher resolution versions of the event using hdfWaterfall.py. Since the bursts are highly circularly polarized try processing the file to Stokes parameters:

```
hdfWaterfall.py -s ##.# -d ##.# -a 0.01 \
```

   has been provided in https://lda10g.alliance.unm.edu/tutorial/Jupiter/. This file has a temporal resolution of 5 s and a spectral resolution of 4.8 kHz. Display this file using:

```
-k 055917_000007389_Jupiter.dat
```

   where "-s" sets the relative offset from the beginning of the file, "-d" specifies the duration to process, "-a" sets the integration time to 10 ms, and "-k" sets the processing mode to Stokes *I*, *Q*, *U*, and *V*. Note that the larger the time selection, the more disk space, processing time, and memory will be required for this and the following step.
3. Plot the higher resolution files with plotHDF.py to get a better view of the burst structure. Depending on the size of the higher resolution file you may need to display a subset of the data.

7.3.3.2 Solar Bursts

Space required: 4.2 GB

1. The data being analyzed here are part of a program to search for prompt emission from gamma ray bursts (GRB). The data are from Fermi Trigger 4053677423 which turned out not to be a GRB but, in fact, a solar burst. The data are taken at a temporal resolution of 40 ms which makes displaying the full file difficult. Start by creating a lower resolution version of the HDF5 file in https://lda10g.alliance.unm.edu/tutorial/Sun/ using the decimateHDF.py script:

   ```
   decimateHDF.py -s 5 20 056601_000024625-waterfall.hdf5
   ```

   This will create a new HDF5 file that has one-fifth the spectral resolution and one-twentieth the temporal resolution which will be more suitable for displaying.
2. Examine the file using the plotHDF.py script and check both tunings for evidence of bursts.

   ```
   plotHDF.py 056601_000024625-waterfall-decim.hdf5
   ```

   For each burst you find, note the time range of the burst relative to the start of the observation. Once again, the plotting script also allows for the color map, stretch, and scale to be interactively adjusted, which may help identify events.
3. For each of the bursts identified, display the burst in the full-resolution data in plotHDF.py. The script supports displaying a subset of the data via the "`-s`" and "`-d`" flags for "skip" and "duration":

   ```
   plotHDF.py -s ## -d ## 056601_000024625-waterfall.hdf5
   ```

7.3.3.3 Giant Pulses from the Crab Pulsar

Space required: 25 GB

1. The data being analyzed here are part of a project to characterize the low-frequency giant pulsed emission (CGPs) from the Crab pulsar. The data consist of tunings at 60 and 76 MHz with a sample rate of 19.6 MS s^{-1}. To begin searching for the CGPs, first convert the first 100 s of data in /data/network/SummerSchool/CGPs/ to an HDF5 file with 16,384 channels and 40 ms integrations:

   ```
   hdfWaterfall.py -a 0.04 -l 16384 -d 100 056598_000003517
   ```

   The large number of channels will improve the dedispersion by reducing the amount of DM smearing per channel.
2. After the file has been converted, display it using plotHDF.py to get an idea of the overall interference environment and to check for any obvious pulses.

The CGPs will appear as curves moving from the upper left to the lower right of the waterfall due to dispersion.

```
plotHDF.py 056598_000003517-waterfall.hdf5
```

Masking RFI will help identify pulses. The calculateSK.py utility provides a simple way to mask RFI based on spectral kurtosis. To create and apply a basic mask run:

```
calculateSK.py -g 056598_000003517-waterfall.hdf5
```

and examine the updated file in plotHDF.py. You can also manually create and edit the mask stored in an HDF5 file with plotHDF.py. To do this use the Mask menu. You may want to try different masking parameters via the `Mask -> Adjust` Masking Parameters dialog, particularly try adjusting the number of sections used for spectral kurtosis. You can also manually adjust the mask by clicking on points to be removed with the right mouse button. After creating an acceptable set of masks save them to the HDF5 file via `File -> Save`.

3. Since the Crab pulsar has a DM of $\approx$56.79 pc cm^{-3} the giant pulses will be more obvious if the data are first incoherently dedispersed and then spectrally averaged. To perform the dedispersion use the dedisperseHDF.py utility:

```
dedisperseHDF.py 56.79 056598_000003517-waterfall.hdf5
```

and then average using decimateHDF.py:

```
decimateHDF.py -s 256 1 056598_000003517-waterfall-DM56.7900.hdf5
```

This will reduce the number of frequency channels to 64.

4. Display the dedispersed and decimated data using plotHDF.py:

```
plotHDF.py 056598_000003517-waterfall-DM56.7900-decim.hdf5
```

The CGPs will appear in the waterfall as horizontal bursts. To more easily identify the CGPs, you may need to turn on bandpassing via the "Bandpass" menu and flag channels that are dominated by RFI.

7.3.4 All-Sky Imaging

7.3.4.1 Detecting Meteor Radio Afterglows in Orville Images

Space required: 4.7 GB

1. The Orville imager is a wideband all-sky imager that creates 100 kHz bandwidth images over a continuous 19.8 MHz bandwidth every 5 s. These images are then averaged down to six channels for long-term archival storage. Plasma trails from meteors can create radio emission observable by the LWA.

Varghese et al. (2021) found meteor radio afterglows (MRAs) in the 198 channel unaveraged data. For this tutorial, we will look at an averaged six channel version available at https://lda10g.alliance.unm.edu/tutorial/MRAs/, which is closer to what is available in Orville archival images.[6]

2. The .oims can be converted into a .fits file using the OIMS2fits.py script:

```
OIMS2fits.py 58956_080000_30.100MHz_49.800MHz-avg.oims
```

3. The resulting multi-extension fits files can be viewed using ds9 with the -mecube option. For now just pick the 48.2 MHz image:

```
ds9 -mecube 58956_080000_48.2MHz.fits
```

   The MRA detection from this dataset was around 8:15 UTC. We can find the image from around that time by selecting `File -> Header` and picking a header and looking for the "DATE-OBS" keyword. Make a note of the time index number of the last image as shown in figure. Hint: If you assume that each file starts at the top of the hour, you can get close to the right index by converting 15 min into seconds then dividing by the duration of a single image.

4. The MRA is not clearly visible in the regular image, so we will create a difference image to find sources that have changes in brightness from one image to the next. We do this by re-running OIMS2fits.py with the `-diff` option:

```
OIMS2fits.py    58956_080000_30.100MHz_49.800MHz-avg.
oims - diff
```

5. Use ds9 to view the difference images that were created (Figure 7.3):

```
ds9 -mecube 58956_080000_48.2MHz-diff.fits
```

   In the difference images we see quite a few scintillating bright sources and possible RFI on the horizon, but at around 8:15 we see a bright source that looks slightly elongated. It appears and then disappears in the following images. This object is the MRA. Make a note of the time index number where the MRA appears.

7.3.4.2 Performing Flux Scale Calibration and Flux Measurement

In this tutorial, we will perform a rudimentary flux calibration in order to measure the flux of the MRA we detected in the previous tutorial. For this tutorial, we will measure the flux of Cygnus A to derive a correction to the flux scaling. A more rigorous approach is given in Varghese et al. (2021) and would involve measuring calibration sources over multiple images in order to discard measurements most affected by scintillation. Alternatively for an example of flux calibrating pulsar observations see Stovall et al. (2015) and Kumar et al. (2025).

[6] https://lda10g.alliance.unm.edu/Orville/

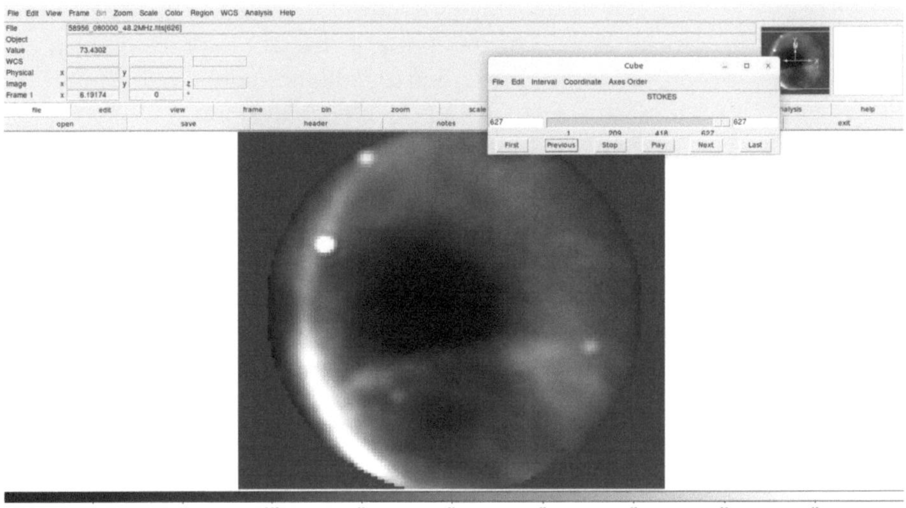

Figure 7.3. The ds9 window is shown with an image opened using the -mecube option. The index of this image is 626 as shown in the field marked "File." Note that the cube window shows the index number+1 because it is one indexed instead of zero-indexed. Credit: Long Wavelength Array Collaboration

1. First we will create a primary beam corrected image of the last image in the 48.2 MHz observation in order to measure the calibrator sources. The primary beam correction will correct for the fall off in sensitivity for sources away from the center of the field-of-view. We can use OIMS2fits.py to perform this correction:

   ```
   OIMS2fits.py -p -i 626 -channel 5 \
   58956_080000_30.100MHz_49.800MHz-avg.oims
   ```

   The `-p` indicates primary beam correction, the `-i 626` indicates that we only want the 626th image in the dataset, and the `-channel 5` indicates that we want the last, highest frequency channel
2. Now we will open the image in ds9 and create a region by selecting `edit -> region.` Click the image and drag the new region into the empties part of the image in order to make a background measurement. In the region window titled "Circle" select `Analysis -> Statistics` and record the RMS measurement. You may want to try moving the window around to see how it affects the RMS number.
3. Edit the location of the region to contain the source Cygnus A which is located at 19:59:28 +40:44:02 (RA DEC J2000). Make the radius of the region 6 degrees.
4. Once again in the region window select `Analysis -> Statistics` and find the Max flux measurement and take a note of it.
5. Find the catalog flux of Cynus A using the VLSS Bright Source Spectral Calculator: https://lda10g.alliance.unm.edu/calspec/calspec.html. Follow the instructions on

the website and use the 3C name along with the frequency to obtain the interpolated flux. Record the given interpolated flux value for Cygnus A. Some of the sources that can be used to calibrate flux from this website include:
   (a) Cassiopeia A (3C 461), 23:23:24 +58:48:54.
   (b) Cygnus A (3C 405), 19:59:28 +40:44:02.
   (c) Crab/Tau A (3C 144), 5:34:30.9 +22:00:53.
   (d) Virgo A (3C 274), 12:30:49 +12:23:28.
6. Using the information we have we will create an averaged correction factor for the flux. This factor can be computed by taking the ratio of the model flux to the background-subtracted flux measurement:

$$corrfac = model/(measured - background) \tag{7.1}$$

   Calculate and record this value.
7. Create a primary beam, flux corrected image of the MRA using OIMS2fits.py with the `-background`, `-corrfac`, `-index`, `-diff`, `-channel`, and `-p` options:

```
OIMS2fits.py -background 32.35 -corrfac 73 -channel 5 \
-diff -i 178 -p 58956_080000_30.100MHz_49.800MHz-avg.oims
```

8. Now we can open the image we create with ds9, create a region around the MRA, and record the flux of the MRA.

7.4 Acquiring Your Own LWA Data

Now that you know how the LWA works, consider using the LWA for a research project of your own. The LWA offers regular calls for proposals in September every year. Typical requests are for 50 beam hours or more, using one or more stations. You can also request smaller allocations at any time through requests to the LWA Director for Director's Discretionary Time (DDT). DDT requests can be made to explore the feasibility of a larger project, or for time-critical observations. In all cases proposals should be sent to `lwa`@unm.edu. The latest call for proposals, along with lots of useful information about the LWA can be found at https://leo.phys.unm.edu/lwa/proposals.html.

References

Hicks, B. C., Paravastu-Dalal, N., Stewart, K. P., et al. 2012, PASP, 124, 1090
Kumar, P., Taylor, G. B., Stovall, K., Dowell, J., & White, S. M. 2025, ApJ, 982, 132
Stovall, K., Ray, P. S., Blythe, J., et al. 2015, ApJ, 808, 156
Taylor, C. A., Dowell, J., Taylor, G. B., et al. 2025, JAI, 14, 2550003
Taylor, G. B., Ellingson, S. W., Kassim, N. E., et al. 2012, JAI, 1, 1250004
Varghese, S. S., Dowell, J., Obenberger, K. S., Taylor, G. B., & Malins, J. 2021, JGRA, 126, e29296

Chapter 8

An Introduction to Solar Radio Bursts

Jason E. Kooi, Joseph F. Helmboldt and George A. Carson

The two strongest sources of radio emission from the sky are the Sun and the background Milky Way Galaxy. The Sun dominates at frequencies above ~300 MHz, while the galaxy dominates frequencies below that. During periods of increased solar activity, though, the Sun can generate greatly enhanced radio emission in the form of solar radio bursts (SRBs) that have a much larger flux density than either the quiescent Sun or the Milky Way. SRBs can emit over large bandwidths in frequency, detectable with ground-based observatories from a few MHz to several GHz. SRBs can cover large timescales, lasting anywhere from seconds to days. SRBs can also generate a range of radio intensities, from just above background levels to orders of magnitude depending on the nature of the emission source (e.g., active regions, solar flares, coronal mass ejections) and emission mechanism (e.g., plasma emission, gyroresonance, electron-cyclotron maser emission). During periods of high solar activity (i.e., "solar maximum," which occurs every 11 years), SRBs can be quite prolific, happening repeatedly throughout the day; however, during periods of low solar activity ("solar minimum"), SRBs may be a rare occurrence. *This chapter includes pure data analysis exercises not requiring equipment.*

8.1 Introduction

The various phenomena associated with SRBs were among the earliest discoveries made during radio astronomy's early years and, no doubt, students will naturally detect SRBs as they engage in the experiments detailed throughout this book. In this chapter, we provide a brief history of their discovery, discuss some of the SRB emission mechanisms and "types" of SRBs commonly observed, as well as provide a guide for accessing SRB data so that students may better understand and identify these events as they embark on their own radioastronomical observations.

doi:10.1088/2514-3433/ae1223ch8

8.2 A Brief History of the Discovery of Solar Radio Bursts

Following the end of World War II, the wartime focus on surface radar systems diminished, leading to laboratories repurposing much of their equipment for research in other fields. It was during this time that radio astronomy began to take off. The discovery that the Sun was a powerful radio source led many scientists to study it (Goss & McGee 2010). One such scientist named Frances Elizabeth Somerville Alexander worked as a Senior Physicist and Head of the Operational Research Section of the Radio Development Laboratory in New Zealand from 1942 to 1945. Despite receiving a PhD in geology from Cambridge University, her main focus in New Zealand was on radio and radar research for the war effort. Her main accomplishments in these areas included pioneering the new field of radio meteorology and developing New Zealand's microwave radar program.

Alexander was also one of the first scientists to observe radio emission from the Sun. Increased background radio noise was detected over Norfolk Island in 1945, dubbed the "Norfolk Island Effect" (Orchiston 2005). To study this phenomenon, she monitored the Sun within an hour of sunrise and sunset with five different radar stations in the area. She concluded that "at sunrise and sunset a detectable amount of noise over and above normal noise is received from a direction roughly that of the Sun" (Alexander 1945). To confirm these results, she planned a large monitoring program in the latter half of 1945, but could not get more than a handful of receiving stations involved. Despite this, she is thought to have observed some of the earliest examples of solar radio bursts with these early radar stations (Harris 2019).

Around the same time, solar radio astronomy also began developing in Australia. Ruby Payne-Scott, employed as a physicist and electrical engineer at the Commonwealth Scientific and Industrial Research Organization (CSIRO) Radiophysics Laboratory, began using radio techniques for astronomical purposes. With the support of her colleague, Joseph Pawsey, she began to observe at a sea cliff with simple radar equipment, which could only record the intensity of the whole Sun at single frequencies. They then began combining observations at different frequencies and noticed the tendency of the brightest emission to drift across the observed frequency spectrum. From this property, they determined that the region of high intensity radio emission must be propagating outward from the Sun (Goss & McGee 2010). This realization inspired the creation of early radiospectrographs, which imaged the whole Sun across many frequencies constantly to detect any bursts and observe their frequency drift rate. However, because early radiospectrographs observed the entire Sun, they did not provide the spatial imaging necessary to pinpoint the source regions for SRBs.

In 1946, Ruby Payne-Scott, Lindsay McCready, and Joseph Pawsey developed early radio interferometry techniques for astronomical observations and were among the first to suggest applying Fourier synthesis to radioastronomical data. These new tools provided the positional information necessary to demonstrate that the SRBs they observed had originated from sunspots (Goss & McGee 2010). This early use of radio interferometry and application of Fourier synthesis not only provided a means to identify sources of SRBs, but also laid the groundwork for aperture synthesis interferometry. Aperture synthesis interferometry involves combining the signals from many small radio telescopes to achieve the same angular resolution as a giant telescope (e.g., the Event Horizon Telescope[1] (Event Horizon Telescope Collaboration et al. 2019) combines signals from a collaboration of telescopes located all around the world to synthesize an aperture the size of the Earth in order to study supermassive black holes at the centers of galaxies).

8.3 Common SRB Emission Mechanisms

Solar radio bursts are classified according to their spatial and temporal characteristics, which are dependent on the nature of the radio emission mechanism generating these bursts as well as the properties of the local plasma. In general, solar radio emission comes in two flavors: coherent and incoherent. Coherent emission mechanisms such as plasma emission and electron-cyclotron emission generate radiation efficiently such that the emitted waves are generally in phase. Incoherent emission mechanisms such as free-free emission (i.e., thermal bremsstrahlung emission), gyroresonance, and gyrosynchrotron emission generate radiation that has no phase association, requiring many particles emitting independently to produce detectable emission (Nindos 2020). The distinguishing characteristic between coherent and incoherent emission is typically the brightness temperature, T_B: if T_B is too large to be associated with incoherent mechanisms, a coherent mechanism is then assumed to interpret the radio emission (Melrose 2009).

Most of the common SRBs we discuss in this chapter are thought to be a result of one of the most ubiquitous forms of coherent radio emission from the Sun: plasma emission. Plasma emission is primarily caused by Langmuir waves, associated with electron density oscillations caused by a displacement of the electrons relative to the positive ions. Following the displacement, the Coulomb force acts to return the electrons to their equilibrium position, similar to the restoring force given by Hooke's law (i.e., $F = -k\Delta x$, where Δx is the displacement and k is the "spring" constant). The electrons, newly imbued with inertia, begin to oscillate around their equilibrium position (Tonks & Langmuir 1929).

Langmuir waves are initiated in the solar corona by a nonthermal electron beam that is accelerated through means such as magnetic reconnection or shockwaves within the plasma. Magnetic reconnection is often tied to solar flare events, and shocks are most often caused by coronal mass ejections (CMEs). The resulting electron beams generate a two-stream instability that, in turn, generates a high-energy tail within the plasma's particle velocity distribution function (sometimes called a "bump-on-tail" instability for this reason). This enhanced velocity distribution tail facilitates transfer of energy from the electron beam to Langmuir wave modes in the ambient coronal plasma. A fraction of the energy contained within the Langmuir waves is then released in the form of radio emission. The fundamental frequency of this emission is the electron plasma frequency, given in Gaussian cgs units as:

[1] https://eventhorizontelescope.org/

$$f_{\rm pe} = \sqrt{\frac{e^2 n_{\rm e}}{\pi m_{\rm e}}} \approx 8980\sqrt{n_{\rm e}} \quad [\text{Hz}] \tag{8.1}$$

where e and $m_{\rm e}$ are the electron charge and the mass of an electron, respectively. The plasma density, $n_{\rm e}$, is given in cm^{-3}. Radio frequencies between 10 and 300 MHz correspond to $n_{\rm e} \sim 10^6$–$10^9\ \text{cm}^{-3}$. Whether the fundamental frequency (Equation 8.1) or harmonics are observed depends on the nature of secondary processes present, e.g., scattering of Langmuir waves by ion sound waves produces emission at $f_{\rm pe}$ and the coalescence of Langmuir waves generates the second harmonic, $2f_{\rm pe}$ (Melrose 2009).

Another fundamental frequency associated with SRBs is the electron-cyclotron frequency (i.e., electron gyrofrequency). This frequency is associated with gyromagnetic emission generated by incoherent emission mechanisms such as gyroresonance and gyrosynchrotron emission, as well as coherent mechanisms like electron-cyclotron maser emission. This is the frequency associated with the motion of an electron that has been injected into a uniform, static magnetic field. The electron will move in a uniform circular pattern with frequency given by

$$f_{\rm ce} = \frac{eB}{2\pi m_{\rm e} c} \approx 2.8B \quad [\text{MHz}] \tag{8.2}$$

where c is the speed of light and B is the magnetic field strength in Gauss. Both Equations (8.1) and (8.2) highlight the basic utility of studying SRBs: their emission can provide insights into the physical characteristics governing the solar atmosphere, most notably the plasma density and magnetic field.

8.4 Common "Types" of Solar Radio Bursts

After their discovery, scientists began categorizing SRBs into separate "types" based on their temporal and spectral structure. Initially, this included only Types I, II, and III (Goss & McGee 2010), but these were later expanded to include Types I, II, III, IV stationary, IV moving, and V, as well as many other highly specialized types (Dulk 1985). A summary of the characteristics for some of the common burst types is given in Table 8.1, adapted from Dulk (1985) and the Australian Space Weather Forecasting Centre.[2] An additional elucidative discussion can be found in White (2024).

Type I bursts are short-lived ($\leqslant$1 s) with little to no frequency drift and tend to occur in "noise storms." Despite individual bursts being short-lived, storms can last hours or even days with their own observed frequency drift as the frequency of each Type I burst slowly changes. An example of a Type I noise storm is shown in Figure 8.1, which has a detectable frequency drift of $-0.3\ \text{MHz s}^{-1}$ with each successive Type I burst. Type I bursts are predominantly observed below 300 MHz and often occur alongside continuum emission. The emission mechanism associated with Type I bursts is fundamental plasma emission (i.e., emitted at $f_{\rm pe}$) and,

[2] https://www.sws.bom.gov.au/World_Data_Centre/1/9/3

Table 8.1. Summary of Characteristics for Common Solar Radio Bursts at Frequencies $\lesssim$1 GHz

Type	Characteristics	Duration	Frequency Range	Associated Phenomena	Emission Mechanism
I	Short, narrow-bandwidth bursts	$\lesssim$1 s	50–350 MHz	Active regions, flares, eruptive prominences	Fundamental plasma
I storm	Consists of high numbers of Type I	Hours to days	50–350 MHz	Active regions, flares, eruptive prominences	fundamental and/or harmonic plasma
II	Slow frequency drift	3–30 min	10–200 MHz	Flares, shockwaves, coronal mass ejections	fundamental and harmonic plasma
III	Fast frequency drift	Single: 1–3 s group: 1–5 min	10 kHz–1 GHz	Active regions, flares, coronal mass ejections	fundamental and harmonic plasma
III storm	Consists of high numbers of Type III	hours to days	10 kHz–1 GHz	Active regions, flares, coronal mass ejections	fundamental and harmonic plasma
IVs	Stationary: broadband continuum, fine structure	hr–days	10 MHz–2 GHz	Flares	Fundamental plasma
IVm	Moving: broadband, slow frequency drift, smooth continuum	min–hr	10 MHz–1 GHz	Eruptive prominences, shockwaves, coronal mass ejections	plasma and/or gyrosynchrotron
V	Short, smooth continuum	1–5 min	10–200 MHz	Often follow Type III rarely occur in isolation	Harmonic plasma

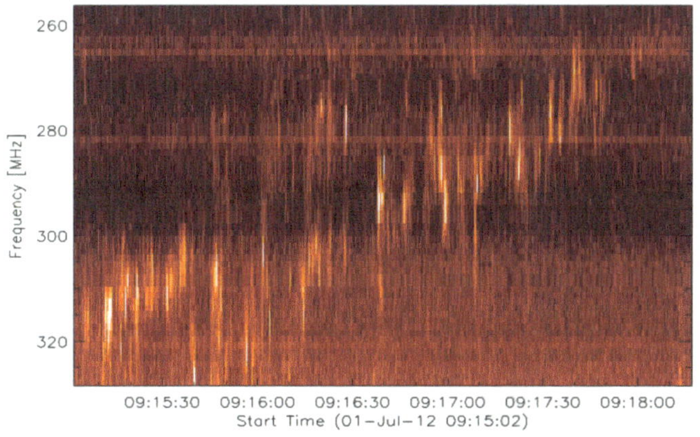

Figure 8.1. Radiospectragraph of a Type I burst storm on 2012 July 1 by e-CALLISTO station in Trieste, Italy. Note that the *y*-axis is flipped so the storm has a frequency drift from high to low frequencies. Reproduced with permission from Karlický (2017). © ESO

therefore, has a large (50%–100%; Dulk 1985) fractional circular polarization. Type I noise storms can indicate energetic processes within closed magnetic loops high in the corona and can also act as precursors to eruption events such as coronal mass ejections (Vourlidas et al. 2020).

Type II bursts are relatively long ($\geqslant$ 10 min) and typically emit at both the fundamental and harmonic plasma frequencies. Unlike Type I bursts, Type II bursts lack a strong circular polarization component, possibly related to dispersion effects. Type II bursts are associated with shockwaves in the ambient plasma and are therefore frequently observed alongside coronal mass ejections that generate a shock front. As this shock front travels outward from the Sun, the electron plasma density in the surrounding solar wind decreases, causing the plasma frequency to slowly drift downward. Because these shock fronts can persist over large regions of the solar wind, Type II bursts tend to cover a broad range of frequencies. As demonstrated in Figure 8.2, the fundamental and harmonic Type II bursts can split into separate bands. The frequency difference between the two bands is related to the shock-compression ratio and, thus, can provide estimates for the Alfvén Mach number that can be used to approximate the strength of the local solar wind's magnetic field (Mahrous et al. 2018).

Type III bursts are fast (individual bursts last $\lesssim$ few seconds) and also emit at fundamental and harmonic plasma frequencies. Examples of Type III bursts, observed with both spacecraft and ground-based stations, are also shown in Figure 8.2. While shockwaves perturb the solar wind plasma to generate Type II bursts, Type III SRBs are generated by electron beams propagating outward along open magnetic field lines within the corona. Similar to Type II bursts, Type III are also categorized by a frequency drift; however, the drift rate is much faster, varying from several MHz s^{-1} to $\lesssim$100 MHz s^{-1} (Miteva et al. 2022). This rapid rate is a result of the speed ($\approx$0.1–0.5 c) at which the electron beams move through the ambient solar wind plasma (Reid 2020). These electron beams are generally

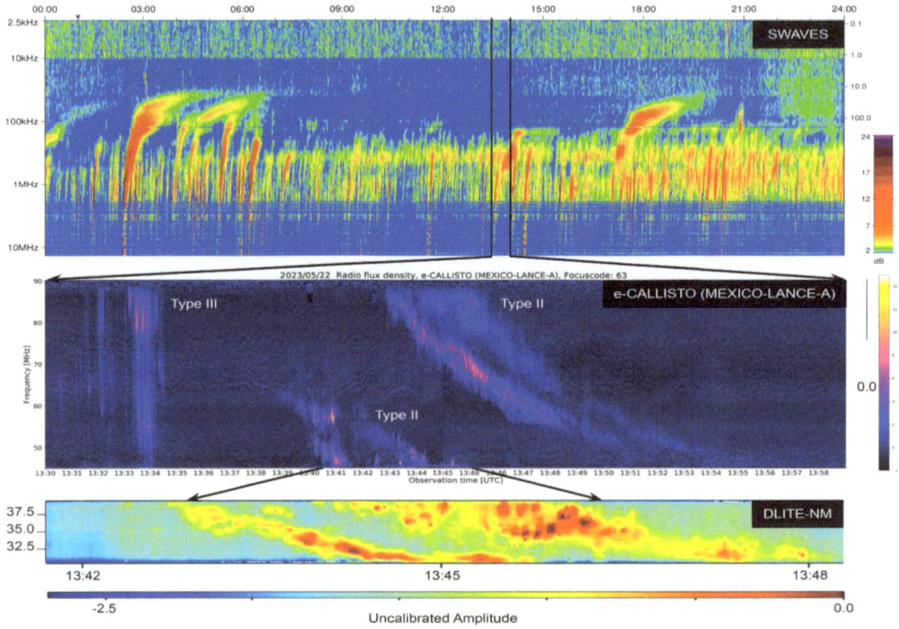

Figure 8.2. On a very active day, the Sun will produce numerous SRBs. The top panel shows spacecraft measurements (STEREO-A/WAVES) of SRBs over the full 24 hr period on 2023 May 22 at frequencies of 2.5 kHz–10 MHz (note the inverted frequency axis). The middle panel provides a higher level of detail from a ground-based e-CALLISTO station (MEXICO-LANCE-A) observing at frequencies of 45–90 MHz. In these data, a Type III radio burst is clearly followed by a band-splitting Type II radio burst. The upper and lower Type II bursts represent the second harmonic and fundamental electron plasma frequencies, respectively. Both the second harmonic and fundamental Type II bursts split into upper and lower bands. The bottom panel shows the fine structure associated with the upper and lower bands of the fundamental plasma emission using a Deployable Low-band Ionosphere and Transient Experiment (DLITE) station in New Mexico, USA, observing at frequencies of 30–40 MHz.

associated with open magnetic field lines and magnetic reconnection events; therefore, Type III SRBs can provide insights into flux tube structures, e.g., see McCauley et al. (2017) and solar energetic particle events, e.g., see Klein (2021). Both individual and groups of Type III bursts can be observed over large frequency bandwidths, providing density information for the solar corona (i.e., Equation (8.1)) over large ranges in heliocentric distance; consequently, they can be used to determine empirical formulas for coronal plasma density, e.g., see Leblanc et al. (1998).

As illustrated in Figure 8.2, Type III SRBs sometimes precede Type II. This is because the same eruption processes (e.g., magnetic reconnection) that generate coronal mass ejections (which, in turn, can generate a shock front that produces a Type II burst) can also release electron beams into the corona that produce Type III SRBs. Similarly, there are several other SRBs that tend to be associated with Type III bursts or are generated by the same underlying principle (i.e., electron beams exciting Langmuir waves within the ambient plasma). Type V bursts are a form of continuum

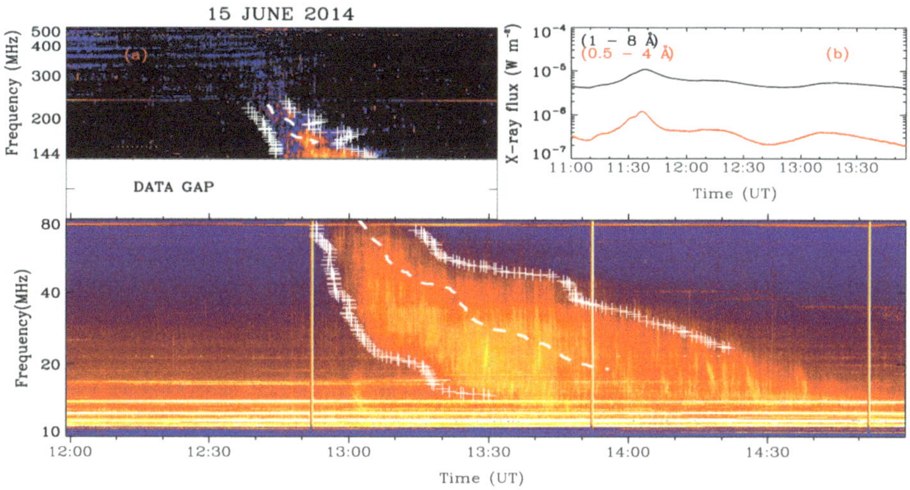

Figure 8.3. Radiospectragraph of a Type IV-moving SRB on 2014 June 15 observed by ORFEES at 144–300 MHz (upper left panel) and the Nançay Decametric Array, NDA, at 10–80 MHz (lower panel) in France. The upper left panel provides corresponding GOES soft X-ray flux data. The white "+" symbols give the upper and lower bounds of the radio emission and the dashed line gives the average frequency. Reproduced from Vasanth et al. (2019). © 2019. The American Astronomical Society. All rights reserved.

emission that are only observed following a Type III burst; consequently, they are likely caused by the same electron beams responsible for the preceding Type III SRBs. Type J and Type U bursts (named for the shape they generate in radiospectrographs) are a subset of Type III bursts in which the frequency drift reverses (from decreasing in frequency to increasing); this is believed to be the result of the electron beam changing from moving outwards into the corona to moving back toward the Sun because it is trapped within a closed magnetic loop.

Type IV SRBs produce broadband continuum emission that can either slowly drift in frequency (Type IV-moving, e.g., Figure 8.3) or appear to remain fixed in frequency (Type IV-stationary); consequently, Type IV bursts can last from tens of minutes to several hours with very little frequency drift. These bursts are believed to originate from nonthermal electrons that are trapped in closed magnetic structures, producing coherent radio emission via plasma or electron-cyclotron maser mechanisms or incoherent emission through gyrosynchrotron radiation. If the particular type of emission can be identified, it can provide insights into the magnetic field structure confining the plasma as well as the nonthermal particle distribution interacting with the field (Carley et al. 2020).

8.5 Examples of Solar Radio Burst Detectors

There are a multitude of solar radio burst detectors active all over the world. Examples of both ground-based and space-based detectors are shown in Figure 8.4, showcasing the wide variety of forms radio telescopes can take: dishes employing reflective plates or wire mesh (e.g., CBSm); crossed-dipole antennas (e.g., DLITE-POM); log-periodic

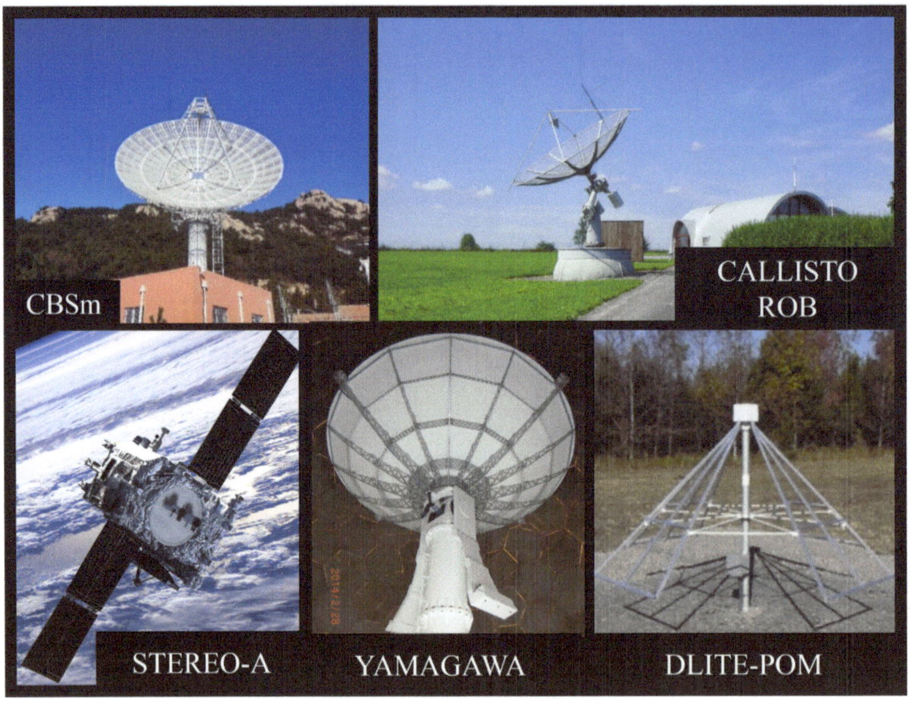

Figure 8.4. Examples of the wide variety of radio instruments located all around the world and beyond capable of detecting solar radio bursts: the CBSm in China; CALLISTO ROB in Belgium; the orbiting STEREO-A; YAMAGAWA in Japan; and DLITE-POM in USA. The CALLISTO ROB instrument uses the log-periodic antenna (indicated by the red arrow) mounted to the side of the 5 m dish. CBSm: Reproduced from Chang et al. (2024). © 2024. The Author(s). Published by the American Astronomical Society. CALLISTO ROB: C. Marqué, Royal Observatory of Belgium. STEREO_A: NASA. YAMAGAWA: Reproduced with permission from NICT. DLITE-POM: Reprinted from Carson et al. (2022). CC BY 4.0.

antennas (e.g., CALLISTO ROB); and monopolar stacer antennas (e.g., onboard the STEREO spacecraft).

The Chashan Broadband Solar radio spectrograph (CBSm; Chang et al. 2024) is a relatively new detector that began observing in November, 2022. It is supported by the Chinese Meridian Space Weather Monitoring Project (Phase II) and Shandong University. It observes a frequency range of 90–600 MHz with a frequency resolution of 76.294 kHz, using a 12 m parabolic dish. It has a time resolution of 0.839 ms, with a maximum configurable time resolution of 0.21 ms, for extremely detailed observations of short-lived SRB structures.

The YAMAGAWA radio spectrograph (Iwai et al. 2017) is operated by the National Institute of Information and Communications Technology (NICT) in Japan. The circularly polarized spectrograph operates with a frequency range of 70 MHz to 9 GHz with a frequency resolution of 31.25 kHz in the MHz regime, and

1 MHz in the GHz regime. It also has an 8 ms time resolution, providing high-resolution observations of shorter bursts such as Type III bursts.

The Deployable Low-band Ionosphere and Transient Experiment (DLITE; Helmboldt et al. 2021) was originally designed by the United States Naval Research Laboratory to study the Earth's ionosphere in detail at frequencies of 30–40 MHz; therefore, it can also be used to detect SRBs (Carson et al. 2022). There are currently four active stations located within the USA (New Mexico, Maryland, Florida, and Ohio), with more stations being built across the world. Each station is comprised of four Long Wavelength Array (LWA; Taylor et al. 2012) antennas that are sensitive to frequencies of 20–80 MHz; however, the current DLITE stations only observe within the range of 30–50 MHz. DLITE stations have frequency and standard temporal resolutions of 16.276 kHz and 1 s, respectively, though DLITE is capable of providing data with a time resolution of 0.1 s. The inexpensive cost and scientific versatility of DLITE make it an ideal array for colleges and universities to incorporate into their facilities, hence its growing popularity around the world.

The Solar TErrestrial RElations Observatory, or STEREO (Karlický 2017), is a set of two nearly identical spacecraft that were launched in 2006 as part of a solar-dedicated space mission; however, since 2014, only STEREO-A is still operational. Included among several solar observing instruments is the STEREO/WAVES (SWAVES) instrument (Bougeret et al. 2008), which supports STEREO's goal to study the 3D structure and evolution of coronal mass ejections by observing Type II and III SRBs. SWAVES uses three mutually orthogonal, 6-meter stacer monopoles that provide SRB data including intensity, source direction, and angular size. Because SWAVES is located in space, it is capable of observing at frequencies of 16 MHz down to 40 kHz, which are typically inaccessible to ground-based instruments due to interference from Earth's ionosphere. Additionally, the SunRISE mission will also be observing the radio sun from orbit (see Chapter 9).

In order to provide continuous, 24 hour coverage of SRBs, an extended network of radiospectrographs has been distributed around the world. These Compact Astronomical Low-frequency, Low-cost Instrument for Spectroscopy in Transportable Observatories (CALLISTO) spectrographs form the e-CALLISTO network (Benz et al. 2005). Different stations within the network employ different types of antennas, ranging from LWA to log-periodic; consequently, there is some variability between each station's operating parameters. The CALLISTO spectrometers themselves operate over a range of 45–870 MHz with frequency and temporal resolutions of 62.5 kHz and 0.25 s, respectively.

8.6 Solar Radio Burst Data Access

Depending on the frequency bandwidth or temporal resolution of the instrument students employ for observing, they may only catch a glimpse of a SRB within their data or, perhaps, all their data may be momentarily saturated by a single SRB or even for hours during a SRB storm. To better understand what they have observed, it can be quite useful to seek out contemporaneous SRB data from one of the multitude of observatories that provide data on a near-daily basis. Many SRB detectors provide their data publicly, ranging from low-resolution quick-look spectrograph images to full-resolution data available in a number of formats (e.g., ASCII, FITS, etc.).

SRB data from an individual, ground-based observatory are usually provided on said observatory's website. Data from the YAMAGAWA detector is located directly on NICT's website: https://solarobs.nict.go.jp/https://solarobs.nict.go.jp/. DLITE data is made available upon request (for more information, see Chapter 10). Data for each e-CALLISTO station is publicly available on their primary website: https://e-callisto.org/Data/data.html. Their data access page includes SRB data in the standardized Flexible Image Transport System (FITS) format from all stations, daily overview spectra, and catalogs of different SRB events detected by each station sorted by date and detector. For those interested in Type II and III SRBs, Hettiarachchi et al. (2024) has developed a Python program to read e-CALLISTO data, integrating burst recognition and analysis tools with a user-friendly graphical interface.

A variety of spacecraft data is provided by NASA's Space Physics Data Facility (SPDF) using the Coordinated Data Analysis Web (CDAWeb) site: https://cdaweb.gsfc.nasa.gov/. This includes SRB data from missions such as STEREO/WAVES, WIND/WAVES (Bale et al. 2008; Bougeret et al. 2008), and the Radio Frequency Spectrometer (RFS; Pulupa et al. 2017), which is part of the FIELDS instrument suite (Bale et al. 2016) onboard Parker Solar Probe (PSP; Fox et al. 2016). On the main page, you can select data by "Source" (e.g., STEREO) or by "Instrument Type." Each Source and Instrument Type can have anywhere from a few entries to >100; consequently, it is best to combine search terms. A practical example of performing this search to generate plots like those in Figure 8.5, which shows a complex set of SRBs from 2024 September 14, is provided in Section 8.8.

Data from the Radio and Plasma Waves (RPW; Maksimovic et al. 2020) instrument onboard Solar Orbiter (Müller et al. 2020 is also available on CDAWeb; however, students may find it easier to navigate the RPW data repository hosted by LIRA, Observatoire de Paris (https://rpw-datacenter.obspm.fr/) or the Radio Monitoring survey project (https://secchirh.obspm.fr/), which also hosts data from other space-based instruments (e.g., STEREO) and ground-based observatories (e.g., ORFEES and NDA, data from which are shown in Figure 8.3). The SunRISE data archive is also publicly available (https://sunrise.umich.edu/data-archive).

8.7 Solar Radio Burst Food for Thought

Whether detecting SRBs is the student's primary goal, or merely understanding the source of the radio loud burst of noise obfuscating observations of galactic emission, below are some points meant to stimulate critical thinking and/or discussion.

8.7.1 Exploring the Detector's Characteristic Values for Key SRB Parameters

- Students should first consider the frequency range of their detector. What is the maximum/minimum frequency? What are the corresponding plasma densities (assuming plasma emission, Equation (8.1)) and magnetic field

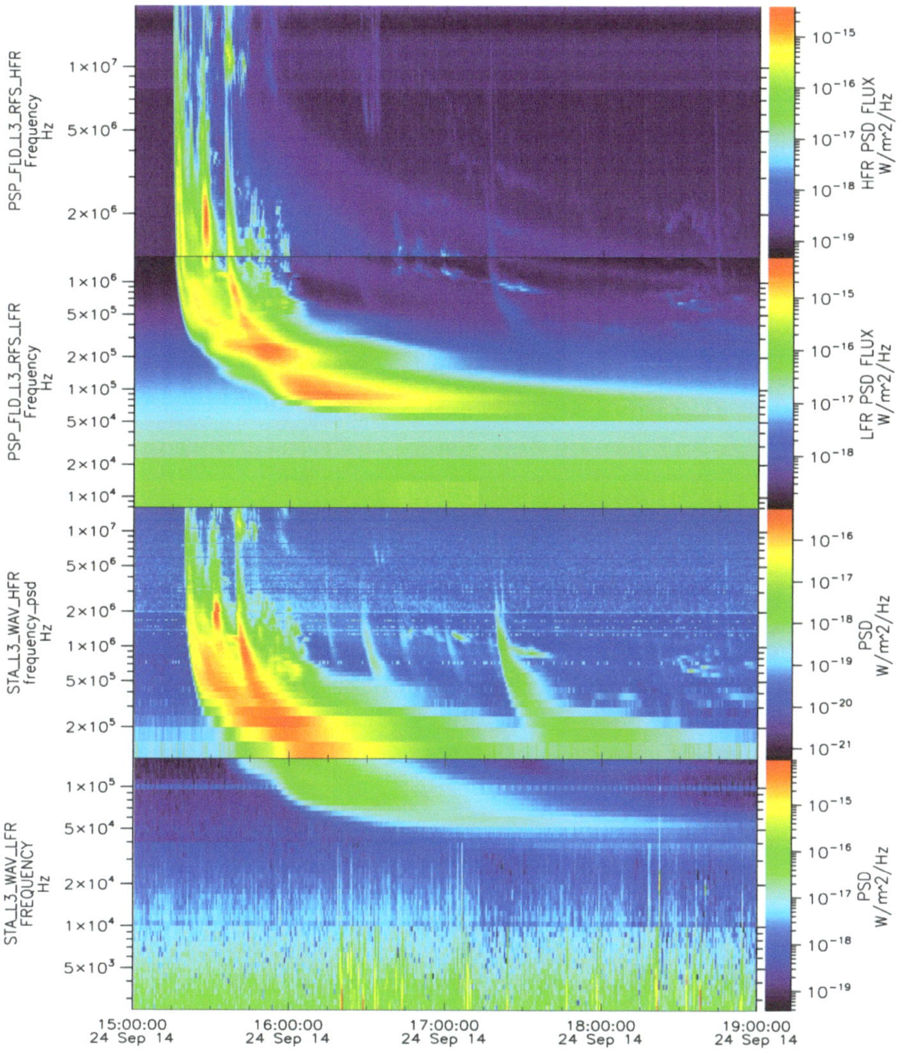

Figure 8.5. 2024 September 14 detection of SRBs by PSP/FIELDS (top two panels) and STEREO-A/SWAVES (bottom two panels).

strengths (assuming emission at the cyclotron frequency, Equation (8.2))? Is the frequency range large enough to capture both fundamental and harmonic emission?

- Knowing the frequency and temporal resolution of the instrument is also important. As mentioned in Section 8.5, a high-frequency resolution reveals more details within the SRB structure and a high time resolution provides access to short-lived bursts such as Type I and III.

- Consider the polarization properties of the instrument. Can it detect circular polarization? If so, students may be able to detect circular polarization from some SRBs, notably gyrosynchrotron emission from Type IV-moving bursts.
- For instruments with narrow bandwidths or low resolution in time or frequency, it can prove beneficial for students to compare their data with data from other detectors; this is discussed in detail through a case study in Section 8.8.

8.7.2 For Radio Interferometers, the Method Used to Detect Sources Can Strongly Impact the Instrument's Sensitivity

As mentioned in Section 8.2, radio interferometers combine the signals of many small antennas to synthesize the angular resolution of a much larger telescope. Interferometers do this using a "correlator" to cross-multiply the voltages (i.e., signals) from the *i*th antenna with the complex conjugate voltages from the *j*th antenna in the array, forming the "visibility" $V_{i,j}$ (for more details, see Chapter 10). Provided the interferometer has at least three elements, then one means of detecting and assessing SRBs is through analysis of the so-called bispectrum (i.e., closure triple). Using the visibilities for three antennas i, j, and k, the bispectrum can be formed by taking the product $V_{i,j}V^*_{i,k}V_{j,k}$, where * represents complex conjugation.

For combinations of three antennas, there are $n_{\text{triple}} = n_{\text{a}}(n_{\text{a}} - 1)(n_{\text{a}} - 2)/6$ closed possibilities for these triple products, where n_{a} is the number of antennas. If the signal-to-noise ratio (S/N) for a single visibility is s, then the S/N for the bispectrum is given by

$$(\text{S/N})_{\text{bisp}} = \frac{1}{2}s^3\sqrt{n_{\text{triple}}} \tag{8.3}$$

Instead of forming the bispectrum, suppose the instrument uses antenna-based incoherent beamforming (i.e., the power detected by each antenna is summed together). The corresponding S/N would be

$$(\text{S/N})_{\text{bfia}} = s\sqrt{n_{\text{a}}} \tag{8.4}$$

The ratio of these two quantities is

$$\frac{(\text{S/N})_{\text{bisp}}}{(\text{S/N})_{\text{bfia}}} = \frac{1}{2}s^2\sqrt{(n_{\text{a}} - 1)(n_{\text{a}} - 2)/6} \tag{8.5}$$

An interferometer like DLITE consists of four antennas ($n_{\text{a}} = 4$), reducing this ratio to $s^2/2$. Provided $s > \sqrt{2}$, the DLITE S/N is larger for the bispectrum and increases rapidly with s. The bottom panel of Figure 8.2 demonstrates a DLITE bispectrum, showing the rich structure of the upper and lower bands for the fundamental plasma emission of a Type II SRB. As Equation (8.5) suggests, it can be highly advantageous for students to compute the bispectrum to analyze SRBs and other bright radio transients if they are using an array of antennas with three or more elements.

For detailed analysis of these and other methods (e.g., coherent beamforming), see Law & Bower (2012).

8.7.3 Discussing the Relative Brightness of Solar Radio Bursts

We asserted that SRBs can be among the brightest radio sources in the sky. It is worthwhile for students to explore just how much brighter these events can be. First, consider generic radio galaxies that are ubiquitous across the sky. A common quantity used to quantify the light received from any given source is spectral flux density, S_ν: the rate at which energy is transferred (i.e., power) from the radio source to a detector, per unit area of the detector facing the source (i.e., projected area), per frequency. In SI units, S_ν is given as W m^{-2} Hz^{-1}. Typical astronomical radio sources are incredibly weak and are given their own unique version of the flux density unit: 1 Jy $\equiv 10^{-26}$ W m^{-2} Hz^{-1}.

The Jy unit is convenient because many galactic and extragalactic radio sources have S_ν ranging from mJy to Jy. "Standard calibrator" sources – so-called because they are used in radioastronomical observations to calibrate the radio data given their large flux densities and well-known, well-understood properties – are typically a few Jy to $\sim$100 Jy at MHz to GHz frequencies. The brightest (extra-solar) sources in the radio sky – sometimes called the "A-Team" sources due to their associations with astronomical constellations (e.g., Virgo A, Cygnus A, Taurus A, etc.) – can be a few 10^3 Jy. Cassiopeia A, the brightest of these sources, has a flux density of $\approx$2700 Jy at 1 GHz (Baars et al. 1977). Nestled inside Taurus A (i.e., the Crab Nebula) is the Crab Pulsar, which has the brightest pulsar pulsations in the Universe at $\sim 10^3$ Jy (Hankins et al. 2003; Cordes et al. 2004).

The quiet Sun (i.e., minimal solar activity) is brighter still. Like the weaker radioastronomical sources described previously, it, too, has its own special flux density unit: solar flux unit, 1 sfu $\equiv 10^{-22}$ W m^{-2} Hz^{-1}. Therefore, even the base unit for solar flux density is 10,000 times larger than the base unit for radioastronomical sources. The quiet Sun generates 10s to 100s of sfu (Tapping 2013). In the hierarchy of flux densities, we finally arrive at solar radio bursts. At frequencies of kHz to GHz, SRBs generate flux densities of 10^2–10^9 sfu, many orders of magnitude greater than even the brightest radio sources.

It is therefore understandable that SRBs can saturate radioastronomical observations, interrupt satellite radio communication and signal transmission, and even interfere with radar systems, e.g., the May 1967 SRB storm that "jammed" the U.S. Department of Defense Ballistic Missile Early Warning System (Knipp et al. 2016). There are a number of ways to avoid the most deleterious effects of SRBs while observing other astronomical sources: software algorithms exist to filter out bright, transient events like SRBs from radio data; observations can be made with directional telescopes (e.g., parabolic dishes) pointed far from the Sun; all-sky arrays can use different beamforming techniques to place the Sun in a null of the beam power pattern. If students are truly interested in detecting weak celestial sources without interruption from SRBs, though, there is a simple trick to avoid these bright events: observe at night.

8.8 Insights into Solar Radio Burst Observations Using Contemporaneous Data: Case Study of 2024 September 14 Coronal Mass Ejection

One of the unique aspects to studying the Sun is that there are numerous facilities, both ground- and space-based, dedicated solely to observing it, providing data across the entire spectrum of light. If students find it difficult determining the nature of the SRB they observed with their own instrument, they should be encouraged to seek out additional observations from other instruments around (or out of) this world. To illustrate this process, we examine contemporaneous data from a number of detectors for an event on 2024 September 14. On this day, several coronal mass ejections erupted, one of which generated an X4.5-class solar flare along with a series of SRBs. This coronal mass ejection hit the Earth two days later, causing a geomagnetic storm with a maximum Kp index of 7.67 (which equates to a moderate strength geomagnetic storm).

Some Type III bursts were captured by ground-based stations with observing frequencies down to 10 MHz; however, most of the rich structure of this complex of SRBs is only detectable below 10 MHz, a domain primarily restricted to spacecraft observations. To observe the full profile, we navigate to CDAWeb to obtain data from the Radio Frequency Spectrometer (RFS) onboard Parker Solar Probe and the SWAVES instrument onboard STEREO-A.

1. On the main page (https://cdaweb.gsfc.nasa.gov/), select:
    - "Parker Solar Probe (PSP)" under the "Select zero OR more Sources" column.
    - "STEREO" under the "Select zero OR more Sources" column.
    - "Radio and Plasma Waves (space)" under the "Select zero OR more Instrument Types" column.
2. Then select "Submit" near the bottom of the list.
3. This will open a new page listing all the "Radio and Plasma Waves (space)" instruments on the "Parker Solar Probe (PSP)" and "STEREO" spacecraft, in order. From the options listed, select:
    - "PSP_FLD_L3_RFS_HFR" – this provides Level 3 data from the high-frequency RFS instrument.
    - "PSP_FLD_L3_RFS_LFR" – this provides Level 3 data from the low-frequency RFS instrument.
    - "STA_L3_WAV_HFR" – this provides Level 3 data from the high-frequency SWAVES instrument onboard STEREO-A.
    - "STA_L3_WAV_LFR" – this provides Level 3 data from the low-frequency SWAVES instrument onboard STEREO-A.
4. Then select "Submit" near the bottom of the list.
5. This will generate a list of ALL data products available for each instrument. From the options listed, select:
    - "Start Time" – "2024/09/14 15:00:00.000."
    - "End Time" – "2024/09/14 19:00:00.000."

- "Plot data" – this is the default option.
- "Also create PS and PDF best quality outputs" under "Plot data" to generate a high quality plot.
- "Combine all time-series and spectrogram plots, for all requested datasets, into one plot file." under "Plot data" to combine the four observational data products below into one plot.
- "HFR Flux Density (Stokes I) data, from cross dipole measurements. (SUMMARY,PRIMARY_VAR)" under the "PSP_FLD_L3_RFS_HFR" instrument heading.
- "LFR Flux Density (Stokes I) data, from cross dipole measurements. (SUMMARY,PRIMARY_VAR)" under the "PSP_FLD_L3_RFS_LFR" instrument heading.
- "Power spectral density after 1% background subtraction and antenna calibration" under both the "STA_L3_WAV_HFR" and "STA_L3_WAV_LFR" instrument headings.

6. Then select "Submit" near the bottom of the list.

Figure 8.5 was generated by following the steps outlined above. This plot showcases the SRBs in their full glory, with more detail than can be seen in ground-based spectrographs due to the low frequency of these emissions. The flux density (provided by the color bars) peaks above 10^{-15} W m^{-2} Hz^{-1}, many orders of magnitude brighter than the brightest astronomical radio sources at these frequencies (see Section 8.7). For more quantitative analysis, the data can also be downloaded in several file formats (CDF, ASCII, CSV). There are many different representations of the data in Figure 8.5 available on CDAWeb. Students can click on the "Info/DOI" link for each instrument to read more details about the instrument, each available data product, as well as see the appropriate citations to the corresponding instrument/mission papers.

A word of caution when using spacecraft data to better understand ground-based observations of SRBs: it is important to consider the position of the spacecraft with respect to the Earth and Sun. If the spacecraft in question is located on the opposite side of the Sun from Earth, it is viewing an entirely different solar environment. Even if its viewing angle is perpendicular to the Sun–Earth line, an SRB may not be detectable even if it is readily visible from the Earth. The SRB on 2024 September 14 was selected in part because of the favorable alignment between the Sun, the Earth, and the spacecraft.

Some spacecraft provide consistent views of the Sun. The Solar and Heliospheric Observatory (SOHO; Domingo et al. 1995) orbits the Sun–Earth Lagrange Point L1, providing continuous observations of the Sun with the same perspective as Earth's. It plays host to a space-based coronagraph called the Large Angle and Spectrometric Coronagraph (LASCO; Brueckner et al. 1995). Coronagraphs are white-light imagers that generate their own artificial solar eclipse in order to study the young solar wind at heliocentric distances of a few to a few tens of solar radii (1 $R_\odot \equiv 695,700$ km at optical frequencies). While this instrument does not provide additional radio data, it does provide unique insights into the plasma dynamics of the solar environment from which these SRBs originate.

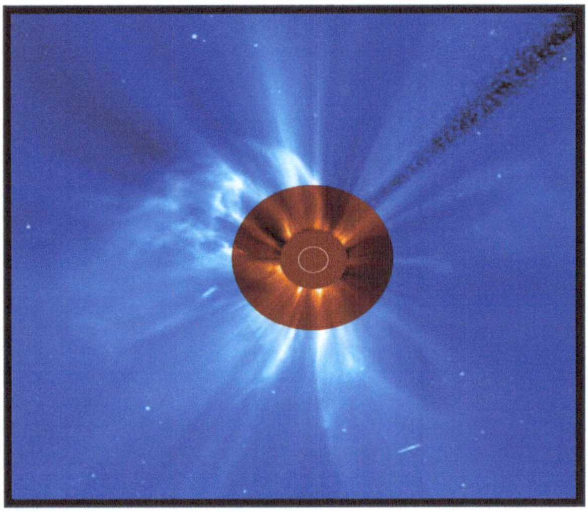

Figure 8.6. Earth-directed coronal mass ejection (CME) captured by the SOHO/LASCO-C2 (red) and LASCO-C3 (blue) on 2024 September 14. The white circle in the center represents the Sun's photosphere. Image Credit: NASA

Figure 8.6 shows an overlay of LASCO images from the C2 and C3 coronagraphs that showcase the Earth-directed coronal mass ejection that generated the sequence of SRBs in Figure 8.5. Coronal mass ejections can display a variety of interesting morphologies. In this case, because the CME erupted toward the Earth, it appears to fan outwards, almost forming a halo around the Sun; thus, these events are sometimes classified as "halo CMEs." Fortunately, generating images like those in Figure 8.6 is relatively simple. Students can navigate to the SOHO homepage at https://soho.nascom.nasa.gov/home.html, click on the "data/archive" tab, and then go to the "SOHO Movie Theater." From here, students can choose an imaging instrument (e.g., LASCO-C2 or LASCO-C3), image resolution (512 or 1024 pixels), and specify the date range. It is important to note, though, that these images have been enhanced to make the features more striking and, therefore, are not suitable for quantitative scientific analysis, unlike the CDAWeb data demonstrated earlier.

Through exploration of contemporaneous data like those available from radio detectors and white-light imagers onboard spacecraft, students may gain a greater appreciation of the SRBs they may observe. In this case study, a ground-based detector operating within the MHz regime might only capture some of the Type III bursts; however, examination of the kHz frequencies accessible to space-based instruments as well as white-light imaging of the solar wind around the Sun tells a much more compelling story of an erupting coronal mass ejection, generating numerous bursts in the lower density (hence, lower frequency) extended corona, as it drives toward Earth. If students are interested in understanding what happens when such coronal mass ejections impact the Earth, students should be encouraged to visit the National Oceanic and Atmospheric Administration's Space Weather Prediction Center (https://www.swpc.noaa.gov/).

References

Alexander, F. E. S. 1945, Report on the investigation of the "Norfolk Island Effect", Technical Report RD 1/518, Radio Development Laboratory 08

Baars, J. W. M., Genzel, R., Pauliny-Toth, I. I. K., & Witzel, A. 1977, A&A, 61, 99

Bale, S. D., Ullrich, R., Goetz, K., et al. 2008, SSRv, 136, 529

Bale, S. D., Goetz, K., Harvey, P. R., et al. 2016, SSRv, 204, 49

Benz, A. O., Monstein, C., & Meyer, H. 2005, SoPh, 226, 143

Bougeret, J. L., Goetz, K., Kaiser, M. L., et al. 2008, SSRv, 136, 487

Brueckner, G. E., Howard, R. A., Koomen, M. J., et al. 1995, SoPh, 162, 357

Carley, E. P., Vilmer, N., & Vourlidas, A. 2020, FrASS, 7, 79

Carson, G., Kooi, J. E., Helmboldt, J. F., et al. 2022, FrASS, 9, 1026455

Chang, S., Wang, B., Lu, G., et al. 2024, ApJS, 272, 21

Cordes, J. M., Bhat, N. D. R., Hankins, T. H., McLaughlin, M. A., & Kern, J. 2004, ApJ, 612, 375

Domingo, V., Fleck, B., & Poland, A. I. 1995, SoPh, 162, 1

Dulk, G. A. 1985, ARA&A, 23, 169

Event Horizon Telescope CollaborationAkiyama, K., Alberdi, A., Alef, W., et al. 2019, ApL, 875, L1

Fox, N. J., Velli, M. C., Bale, S. D., et al. 2016, SSRv, 204, 7

Goss, W. M., & McGee, R. X. 2010, Under the Radar: The First Woman in Radio Astronomy: Ruby Payne-Scott (Heidelberg: Springer)

Hankins, T. H., Kern, J. S., Weatherall, J. C., & Eilek, J. A. 2003, Natur, 422, 141

Harris, M. 2019, Rocks, Radio and Radar: The Extraordinary Scientific, Social and Military Life of Elizabeth Alexander (London: World Scientific)

Helmboldt, J. F., Markowski, B. B., Bonanno, D. J., et al. 2021, RaSc, 56, e07298

Hettiarachchi, Y., Adassuriya, J., Jayaratne, C., Jayawardhana, S., & Monstein, C. 2024, NewA, 109, 102194

Iwai, Y., Kubo, J., Ishibashi, C., et al. 2017, EP&S, 69, 95

Kaiser, M. L., Kucera, T. A., Davila, J. M., et al. 2008, SSRv, 136, 5

Karlický, M. 2017, A&A, 602, A122

Klein, K. L. 2021, FrASS, 7, 105

Knipp, D. J., Ramsay, A. C., Beard, E. D., et al. 2016, SpWea, 14, 614

Law, C. J., & Bower, G. C. 2012, ApJ, 749, 143

Leblanc, Y., Dulk, G. A., & Bougeret, J.-L. 1998, SoPh, 183, 165

Mahrous, A., Alielden, K., Vršnak, B., & Youssef, M. 2018, JASTP, 172, 75

Maksimovic, M., Bale, S. D., Chust, T., et al. 2020, A&A, 642, A12

McCauley, P. I., Cairns, I. H., Morgan, J., et al. 2017, ApJ, 851, 151

Melrose, D. B. 2009, Universal Heliophysical Processes, Proc. IAU Symp. 257 ed. N. Gopalswamy, & D. F. Webb (Cambridge: Cambridge Univ. Press) 305

Miteva, R., Samwel, S. W., & Zabunov, S. 2022, Univ, 8, 275

Müller, D., St. Cyr, O. C., Zouganelis, I., et al. 2020, A&A, 642, A1

Nindos, A. 2020, FrASS, 7, 57

Orchiston, W. 2005, The New Astronomy: Opening the Electromagnetic Window and Expanding our View of Planet Earth, ed. W. Orchiston (Dordrecht: Springer) 71

Pulupa, M., Bale, S. D., Bonnell, J. W., et al. 2017, JGRA, 122, 2836
Reid, H. A. S. 2020, FrASS, 7, 56
Tapping, K. F. 2013, SpWea, 11, 394
Taylor, G. B., Ellingson, S. W., Kassim, N. E., et al. 2012, JAI, 1, 1250004
Tonks, L., & Langmuir, I. 1929, PhRv, 33, 195
Vasanth, V., Chen, Y., Lv, M., et al. 2019, ApJ, 870, 30
Vourlidas, A., Carley, E. P., & Vilmer, N. 2020, FrASS, 7, 43
White, S. M. 2024, arXiv:2405.00959

Chapter 9

The Sun Radio Interferometer Space Experiment and Student-led Ground-based Observations

Mojtaba Akhavan-Tafti and Shirsh Lata Soni

We present the setup and procedure for our LWA-antenna-based SunRISE Radio Ground Lab, designed for students to collect a complementary ground-based dataset to the space-based CubeSat mission SunRISE, which will observe the Sun at radio frequencies (<10 MHz) inaccessible from the surface of the Earth. One of the main goals of our LWA observing campaign is to investigate and determine the source regions of solar radio bursts. They are characterized by their transient and intense outbursts of radio emissions, commonly associated with solar flares and CMEs. The underlying mechanisms responsible for the generation and propagation of these bursts can be determined by using open source data and employing simple observational techniques and data analysis methods. We show setup and data collection procedure, which will be helpful for any LWA antenna-based project. *Estimated Cost: $25,000 USD*

9.1 Introduction

The Solar Radio Interferometer Space Experiment (SunRISE) mission is a space-based low-frequency radio interferometer designed to explore and understand the characteristics of solar eruptions (Kasper et al. 2024). SunRISE uses an array of six CubeSats operating in supersynchronous geosynchronous Earth orbit (GEO) with a tight separation of 10 km between each satellite (Figure 9.1). This special array is designed to capture detailed images of the Sun and provide important data on the mechanisms that drive particle acceleration associated with solar activities such as solar flares, coronal mass ejections (CMEs), and active regions etc. At the time of this writing, SunRISE is near launch. After launch, an opportunity will emerge for ground-based radio observatories at LWA antenna frequencies (10–88 MHz) simultaneous with space-based SunRISE observations (0.1–23 MHz).

doi:10.1088/2514-3433/ae1223ch9

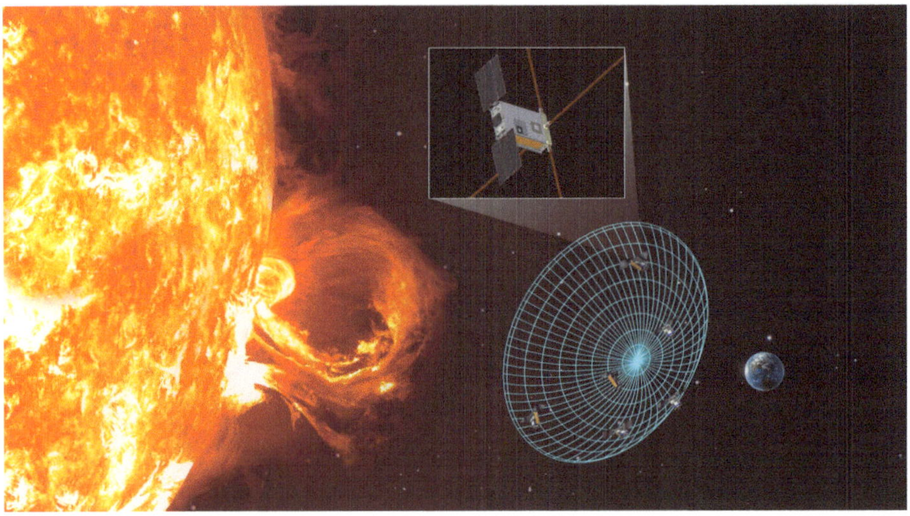

Figure 9.1. Schematic illustration of SunRISE CubeSats' orientation and position (used from https://www.jpl.nasa.gov/missions/sun-radio-interferometer-space-experiment/). Image Credit: NASA

9.2 The Student-Built Long Wavelength Array (LWA) Antenna

9.2.1 Objectives and Goals of the LWA Array in the Context of SunRISE

The LWA antenna setup used here works on a narrow frequency range centered around 20.1 MHz (with a temporal resolution of 1 s and a spectral resolution of 100 kHz), and designed specially to receive the radio emissions from the Sun.

Both the space-based and ground-based components of SunRISE we seek to identify the physical processes and regions within the solar atmosphere where these radio bursts originate. A comprehensive understanding of their source regions is pivotal for advancing our knowledge of solar activity, space weather forecasting, and its potential impact on various technological systems, including satellites and communication networks.

Another important goal of our LWA observations is to detect and characterize ionospheric disturbances induced by solar radio bursts. Radio waves released from solar flares and CMEs known for their intense and transient radio emissions, have the potential to impact the Earth's ionosphere, leading to perturbations in the ionospheric electron density. These perturbations can be identified and quantified by radio wave monitoring with LWA observations and SunRISE. Spaceborne observation of solar radio bursts avoids the ionospheric interference and provides a baseline measurement of the bursts' characteristics. Ground-based measurements at LWA frequencies receive the radio waves after they have passed through the ionosphere. Ionospheric disturbances, such as increased ionization due to solar activity or geomagnetic storms, can affect the propagation of radio waves through the ionosphere. This can lead to signal delays, phase shifts, and amplitude variations in the received radio signals. The disturbances in the received signals can be identified by comparing and contrasting spaceborne and ground-based observations. The ionospheric conditions can then be inferred from the identified radio signal disturbances. This information is essential for understanding how solar activity impacts Earth's ionosphere and can affect communication and navigation systems. An extensive discussion of LWA solar radio observations in the SunRISE context can be found in Akhavan-Tafti et al. (2025). (There, additional student observations with a Radio Jove-like telescope, similar to that in Chapter 6, are also presented.)

9.2.2 Overview of the Long Wavelength Array (LWA)

9.2.2.1 Antenna Specification

The LWA antenna is an inverted-V, tied fork design receiver with four triangular slop wings. We focused on developing a lower cost and higher-performance antenna than the model available for purchase. The frequency range, 5–90 MHz, the dimensions of the dipoles and triangular-shaped wings were obtained using antenna theory. By taking the average frequency as 50 MHz, the wavelength was determined as 5.995 m. Antenna lengths were calculated as nearly 2.58 m. Each of the two dipoles consists of two triangular aluminum frame elements sloping down 45 degrees and supported from the steel center post by non-conductive horizontal fiberglass rod braces. Each element is 1.5 m long × 0.8 m wide. At the base of the antenna is a 3 m × 3 m galvanized wire mesh ground screen. The peak (apex) of the two dipoles is 1.5 m above ground level with the dual active balun assembly in the gray enclosure at the top. Two coaxial cables can be seen leading away from the antenna, one for each dipole (Nelson 2014).

The LWA spectrometer is a receiver instrument, which is designed by Reeve Radio Observatories.

The SunRISE LWA antennas observe solar radio frequencies with its integrated antenna system and uploads the spectra to the main server in the University of Michigan, MI, USA. Observations of solar radio bursts are made using antennas and the radio frequencies are recorded in intensity. The radio spectrographs consist of a dynamic spectrum which is continuously recording intensities of radio flux over a range of frequencies. To set up data collection on a computer, the e-Callisto software is utilized.

9.2.2.2 Observing Site

Our aim is to deploy two antenna arrays, one at Peach Mountain (Figure 9.2) and second antenna array at Michigan's Upper Peninsula, Marquette High School, Marquette, MI. The scientific reason behind installing radio antenna arrays at two locations was to increase the latitudinal coverage of the arrays. The deployments at Peach Mountain and Marquette are 20 (latitudinal) degrees apart, allowing us to study ionospheric disturbances, only observable in higher latitudes, such as auroral activities. The high school was selected based on existing contacts in the area with accessible power and internet.

Figure 9.2. LWA antenna installed at Peach Mountain Observatory, Dexter, Michigan. Diagram of imaged LWA antenna. Included is the ground stake, not visible in the image which provides stability for the antenna against wind and grounding for the antenna.

Our home installation site is Peach Mountain Radio Observatory, located in Dexter, MI about 12 miles (20 km) northwest of the University of Michigan. The observatory was initially constructed in 1955 with support from the United States Navy Office of Naval Research. The site is home to the 85' Peach Mountain Radio Telescope, which opened in 1958 and was used up until 2010 to observe the linear polarization and flux density values for emissions in radio frequencies coming from objects outside of the Milky Way. The site is still frequently used for astronomical research by the University of Michigan.

Peach Mountain Radio Observatory is considered an ideal site for radio astronomy within mainland Michigan due to its isolation from residential and commercial infrastructure, and as such, extraneous sources of noise in radio frequency. The site is gated in multiple locations, which prevents unauthorized vehicles from approaching the site, and all unwanted visitors from entering. As such, long term observation is possible without constant supervision of the instrumentation. For our installation, we required four flat 10' $\times$ 10' plots of land located in relatively close proximity to two power supplies so that the enclosure and antennas could be powered and operate as intended. Staking out of the site was done in advance to choose the best locations for installation. A detailed diagram showing the resulting chosen installation locations is shown in Figure 9.3.

Marquette Senior High School is a public, co-educational high school located in Marquette, MI, serving grades 9–12. The deployment in Marquette helps to expand the beneficiaries of the science project to also include high school students. Faculty from the University of Michigan with the school's science teacher will help students to install the antenna array on campus, collect radio data, and analyze radio data to accomplish our science objectives.

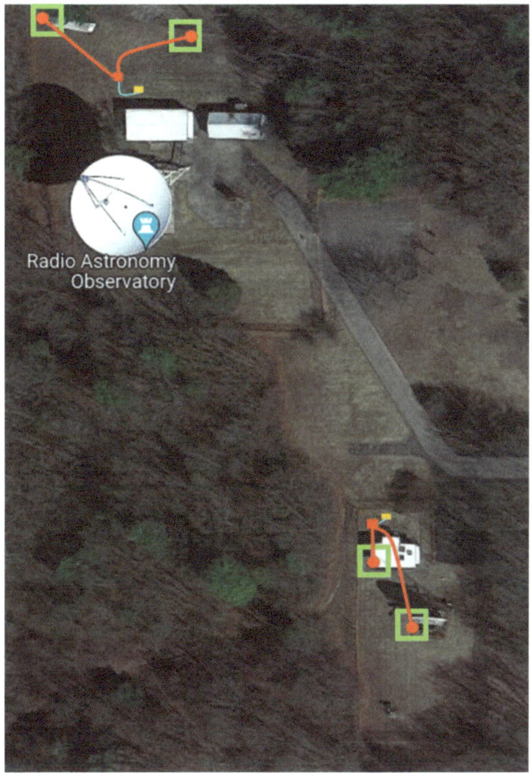

Figure 9.3. Peach Mountain Radio Observatory. The altitude of the site is 1033′ (315 m). The coordinates are 42.3988°N 83.9357°W. A bird's eye view of our installation locations on Peach Mountain. The green boxes show the locations of each ground mesh with the red dots representing the location of each LWA antenna. The solid red lines show the paths of the LMR-400 coaxial cables, buried under the ground within a flexible PVC pipe. The red boxes show the location of the two enclosures. The blue lines show the paths of the 10 AWG Teck 90 power cables and the yellow boxes show the locations of the power outlets. Map data ©2025 Google.

9.2.3 Design, Fabrication, and Technical Specifications of LWA Array

In addition to the SunRISE team's assembly procedure shown here, one should also be familiar with the LWA Memo Series before beginning, especially Memo #225: https://leo.phys.unm.edu/lwa/memos/

9.2.3.1 System Requirements and Installation Instructions

For successful LWA antennas installation, the following parts first need to be acquired: a ground stake, the antenna itself, an LWA PC (Power Coupler), two hydrodyne upconverters and two Callisto Receivers. Fabrication of these components from scratch is possible, but for the time and cost required, we have found that it is much easier to purchase the components. Each of these can be purchased from several possible vendors. The prices[1] for each of these components is shown above in Figure 9.4 and each of these components is shown below in Figure 9.5. For SunRISE, LWA antennas were purchased from

[1] https://reeve.com/RadioScience/Antennas/ActiveCrossed-Dipole/LWA_OrderInfo.htm

Item	Description	Price
LWA-FEE	Front-End Electronics, Active Balun for one crossed-dipole antenna (V2.0 dual assembly)	US $472.95
LWA-ANT	Crossed-dipole antenna including mast	US $792.75
LWA-STK	Ground stake for Item LWA-ANT	US $65.50
LWA-SYS	Includes 1 each LWA-FEE, LWA-ANT, LWA-STK	US $1330.00
LWAPC	LWA Power Coupler without quadrature coupler	US $495.00
LWAPC-Q	LWA Power Coupler with quadrature coupler	US $745.00
SHIP	Shipping	Determined at time of shipment

Figure 9.4. ReeveLWA Installation Part Prices. Radio antennas can be costly, which is why most research using them is conducted through academic institutions throughout the world.

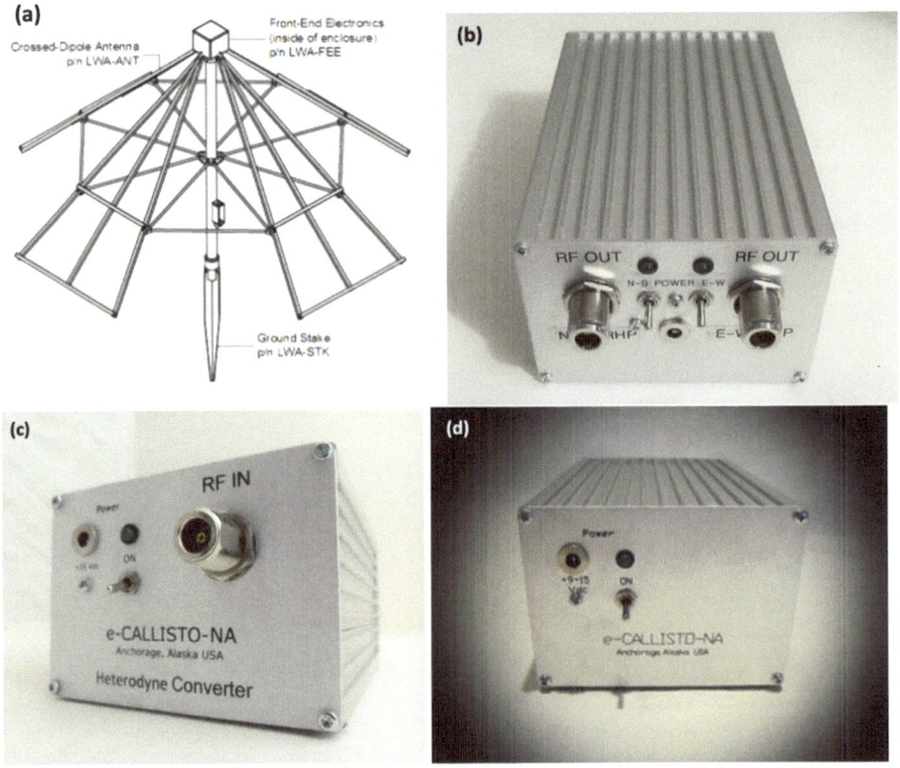

Figure 9.5. LWA Components. (a) Fully assembled antenna. The mast is mounted on the ground stake which is submerged in the ground when fully installed. The Front-End Electronics board shown at the top of the antenna is the location from which the data is able to be collected. (b)LWA Power Converter. This is powered by a regulated 15–17 V DC power supply. It has two input and two output coaxial cable ports. This is because of the polarized signal, where North–South and East–West data are both collected, and then later combined. (c) Upconverter. This is powered by a 15–17 V regulated DC power supply. Depending on the year in which the upconverter was manufactured, it could either have one port for input and one for output, or two of each. If an upconverter has only one of each, then two upconverters are required for installation. Otherwise, one will suffice. (d) Callisto Receiver. Each receiver is powered by a 12 V Nominal DC power supply and has one coaxial cable input and a number of possible outputs. The most commonly used one is the COM port. Image credit: Whitham D. Reeve.

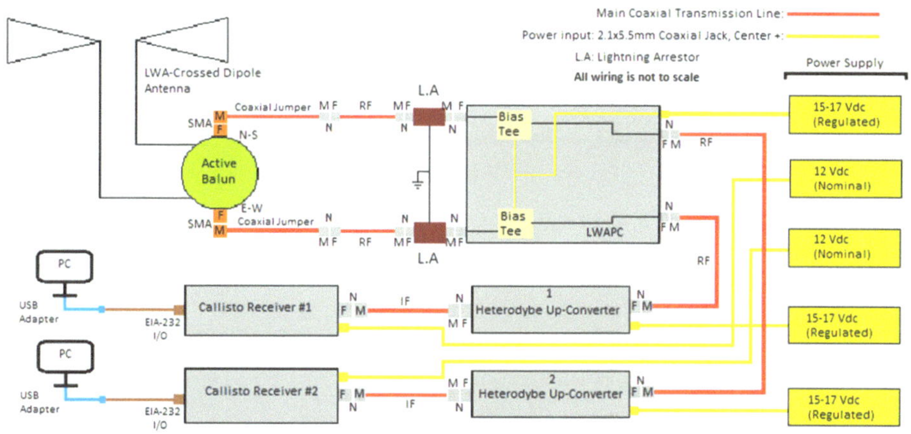

Figure 9.6. System Setup Diagram. Flow chart containing components purchased from Reeve Radio Observatories, their power supplies, and their connection to the computers, which will be discussed shortly. Of note is that this diagram shows two upconverters. However, if one purchases a newer model with two input and two output ports, only one is required. If this is the case make sure N–S and E–W cables stay consistent. The output port directly opposite a given input is the one that it correlates with.

Reeve Radio Observatories; as discussed in Chapter 11, these may also be purchased from Burns Industries.

The connections between each of these components, the power supply requirements, and the cables used to connect the components are all shown below in Figure 9.6.

In addition, several other parts are required such as (1) a weather resistant, temperature controlled enclosure to hold the power coupler, (2) upconverters, (3) receivers, (4) a ground mesh to be installed underneath the antenna, (5) power cables to power each of the components, (6) coaxial cables to transmit data from the antenna, and (7) computers to collect the antenna data. If the computer you choose to use does not have a COM port, USB-Serial adapters are also required.

The coaxial cables used are LMR-400 Coax Cables and the power cables used were 10 AWG Teck 90 Cables. The lengths required for each type of cable is completely dependent on the location of installation. The length of the power cable should be slightly longer than the distance from the nearest power supply to the enclosure. The length of the coaxial cables should be slightly longer than the distance from the enclosure to the mast of antenna where the opening exists (small rectangular box in Figure 9.6a). Coaxial cables are also required for connection between the instrumentation within the enclosure, LWAPC to upconverter(s), upconverter(s) to Callisto receivers. The lengths of these cables is dependent on the setup of enclosure, but typically cables ~1 ft long for each connection should suffice.

The ground mesh is a $10' \times 10'$ mesh of $2'' \times 4''$ 14 AWG Galvanized Steel. Instead, in our case, we found to be most successful was to purchase a $4' \times 50'$ garden mesh from Home Depot, a 50 Pack of $3/32''$ Aluminum Wire Rope

Figure 9.7. Cut up 4′ × 10′ garden mesh. Be careful when cutting. The resulting points created from the cut probes are very sharp. Using cut resistant gloves is much advised.

Compression Sleeves from McMaster Carr (3896T32), and a Multi Diameter Compression Tool (compression tool should be compatible with 3/32″ aluminum sleeves). From there we would use a wire cutter to cut out three 4′ × 10′ segments from the garden mesh. After that, we used a wire cutter to cut away one of the 10′ edges from each of the segments. Cut as close to the edge as possible for every 2 of 3 probes, and 1/2″ up from midway (toward the edge you are cutting) on every 1 of 3 probes, starting from one of the probes on the 4′ edge. If successful, both the first and last probe should be shorter, the resulting component looks like Figure 9.7.

Choose one of the three segments. Measure out 2′ starting from where the cut probes contact their first cross beam in the direction toward remaining intact 10′ edge. After reaching 2′, perform the same cuts previously conducted, such that you are left with a symmetrical 2′ × 10′ garden mesh with probes sticking out on both 10′ edges as shown in Figure 9.7.

Next, take one of the 4′ × 10′ segments and align the cut up 10′ edge with one of the 10′ edges on the 2′ × 10′ segment. If all the tasks were performed successfully, the shortest probes from each of the edges should overlap by approximately 1″. Take one of the 3/32″ aluminum splicing sleeves and run it through one of the short probes, and then back through the paired short probe, such that the sleeve is then centrally located. It may be useful to have someone hold the two components steady to make sure that they do not shift with respect to each other. Using the Compression Tool, and the designated sizing (look for 3/32″ in the manual that will come with your compression tool, likely G), splice the sleeve. If designated in the manual, you may need to splice twice for each sleeve. Repeat this process for each of the short probe pairs. If done correctly, there should be 21 short pairs. Take the

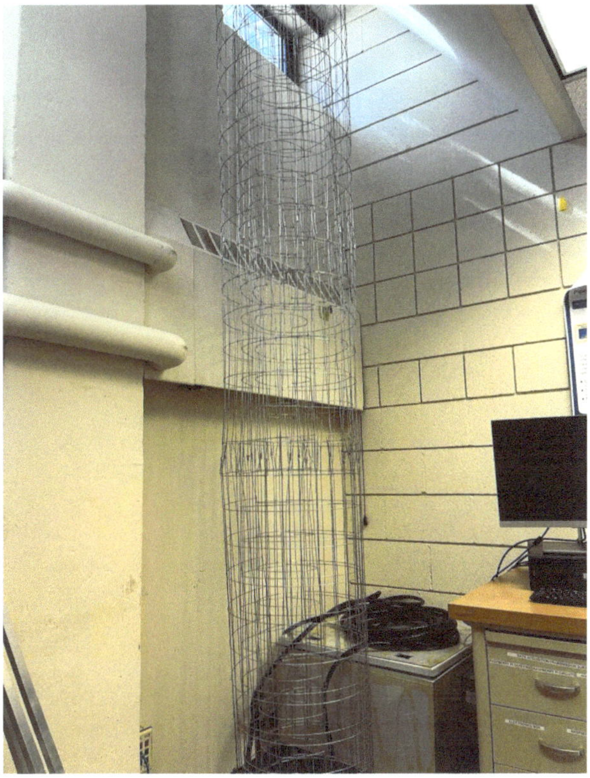

Figure 9.8. Completed Ground Mesh. For ease of storage, you can either roll up the mesh as is shown, or fold it up.

remaining 40 free long pairs and manually twist them around each other. Repeat the entire splicing process again to connect the other side of the $2' \times 10'$ segment and the remaining $4' \times 10'$ segment. This will likely be a big forearm workout so feel free to switch out who is splicing and who is holding the segments. The final ground mesh should look something like Figure 9.8.

For the enclosure system, the most important factors are to make sure that the enclosure is fully weatherproof (rain, snow, wind, hail), thermally stable at 67–71 degrees Fahrenheit and free of unwanted radio interference. Holes must be cut to allow two coaxial cables and one power cable in. These holes must be weatherproof as well. There are many ways to make sure the enclosure is properly weatherproofed and thermally stable. The components we chose to purchase are shown in Figure 9.9.

The last major component required for data collection are the computers. Each antenna will result in two COM cables, one for each polarization state (N–S, E*-W). Computer does not need to have a COM port; a USB port and a USB-Serial adapter combination will act exactly the same. Make sure that the USB-Serial Adapter is compatible with the operating system of chosen computer. In addition, recognize that the chosen computer will have to stay within the enclosure for the entirety of data collection. We selected the Beelink Mini S12, which had enough USB ports to be used for both polarization states.

Subsystem	Item	Item Code	Distributor	Quantity	Unit Price	Total Price	Lead Time
Enclosure	Hoffman Enclosure CSD363616	CSD363616	RSP Supply	2	$1,133.02	$2,266.04	9-10 Days
	Dielectric Grease	2YKJ9	Grainger	2	$43.52	$87.04	2 Days
	Silicone Sealant	49LX63	Grainger	2	$10.15	$20.30	2 Days
Radio Interference	Block Breathable Pure Silver Fiber Mesh Fabric	B0BVR295X4	Amazon	1	$40.38	$40.38	2-3 Weeks
Thermal Control	Polycarbonate Insulation	N/A	EplastUSA	Custom	Custom	$322.56	3 Weeks
	ThermoTECTM Series - 800-950 BTU Thermoelectric Solid State Air Conditioner	AAC-141A-4XT-HC	EIC Solutions	1	$2,440.00	$2,440.00	14-16 Weeks
	1 Gallon Reflective Paint		Exterior Coatings	2	$79.00	$158.00	2 Days
	Hydrosorbent Reusable Slica Gel Desiccant Dehumidifier (5 Pack)	OSG-40	Amazon	1	$49.00	$49.00	2 Days
	Condensate Drip Pan	DP-V	EIC Solutions	1	$85.00	$85.00	14-16 Weeks

Figure 9.9. Purchased Enclosure Components. The name of components, their item code, distributor, quantity, unit price, total price, and required lead time are all shown. The polycarbonate insulation is sized based on the surface area of the chosen enclosure. There should be one panel for each of the sides excluding the bottom. The panel on the two should extend an extra distance such that it covers the extra thickness added by adhering insulation to the outside of each of the four sides.

Figure 9.10. An installed ground mesh. Each rectangular element visible is 2″ × 4″ and pointed in the same direction as the frame elements of the LWA antenna. The yellow vehicles shown are the plastic stakes used to secure the ground mesh to the ground.

9.2.3.2 Ground Screen

A 14 AWG galvanized steel wire mesh is used to create a 10′ × 10′ ground screen for the ground surface surrounding the LWA antenna. The purpose of the ground screen is to reflect radio waves and to remove noise. It does this by circularly polarizing the signal in the same direction of the central bar in each of the four triangular frame elements. For more information regarding circular polarization and why it is required for LWA antennas (and most other antennas) see:

https://reeve.com/Documents/Articles%20Papers/Reeve_Polarization.pdf

To best polarize the signal, the ground on which the screen is placed must be flat and the ground screen must lay flat to the ground. To assure that this is the case, plastic yellow tent stakes are used to secure parts of the ground screen to the ground. These stakes do not interfere with polarization being performed by the ground screen. Shown in Figure 9.10 is an image of an installed ground screen.

9.2.3.3 Data Pipeline and Handling

The Long Wavelength Array (LWA) receives data in a format known as FITS (Flexible Image Transport System), a widely used digital file format in astronomy. This data is processed and analyzed using the astropy library in Python, a powerful tool for handling astronomical data. The FITS files from the LWA contain intricate details about the frequencies detected at specific times, encapsulating essential information about the observed celestial phenomena. Through a process of transformation and analysis, this raw data is converted into a data matrix. This matrix is a structured representation of the emission data, setting the stage for further analysis and visualization, particularly the creation of spectrograms.

The conversion of emission data to a spectrogram involves several crucial steps. The first step is finding areas of transience using a threshold value. This is achieved by scanning the data matrix to locate regions where the emission data's frequency exceeds a certain threshold. This threshold is typically determined by calculating the standard deviation across the entire frequency range, aiding in the identification of significant outliers or unusual emission events. These areas of transience are pivotal as they can indicate unique or noteworthy astronomical events.

Subsequently, the process involves identifying clusters within these transient locations. Clusters are groups of data points with close spatial proximity, suggesting a relationship or common origin among the emission events they represent. This clustering is a critical step as it helps in isolating related events from the vast expanse of data. The final step in the visualization process is creating a binned version of the data. In this step, each bin indicates whether the emission in that region is above the predetermined threshold. This approach effectively creates a heatmap-like representation in the spectrogram, where colors vary based on emission intensity. The resulting spectrogram is a powerful tool for visualizing the frequency and time data from the LWA emissions, highlighting areas of significant activity and facilitating a deeper understanding of the observed astronomical phenomena.

9.2.4 Practice Session

9.2.4.1 Setting Up Antenna and Connecting to the Receiver

Before beginning Installation of the LWA antenna, it may be useful to have a set of (at least 12) plastic tent stakes. These will be used to secure the ground stake in place. In addition, a wood-handled steel post hole digger will be useful in digging the hole for ground stakes. This is not required, but it is recommended as it saves a lot of time and effort, especially in colder weather. The coaxial cables, which go from the antenna to the enclosure, are advertised to be weatherproof, but to be safe, some form of insulation for the coaxial cables is advisable. This could be a flexible PVC tube, burying the cables underground, or a combination of the two. With all that in place installation is ready to commence.

- **Step 1** Find the flattest $10' \times 10'$ region you can which is free from any nearby overhangs (either buildings, trees, or mountains) which would block the Sun and as far from any sources of potential radio interference (primarily commercial and residential buildings) as is possible. Lay the ground mesh

Figure 9.11. Ground mesh, stake and mast installation.

flat on the ground and secure it down using the plastic tent stakes. Find the center of your ground mesh and use wire cutters to cut away a square large enough for the ground stake of the antenna to fit through without contacting the mesh. Within that square, use the post hole digger (or any tools of your choice) to dig out a hole deep enough such that the spokes of ground stake can be fully submerged into the ground. Place the ground stake into the hole and fill in any air pockets with dirt, making sure to keep the ground stake oriented directly upwards. Once the ground stake is secured in the ground, insert the antenna mast into the stake. If successful, the resulting antenna will look like Figure 9.11 below:

- **Step 2** Place FEE support (Figure 9.12) on top of the mast.
- **Step 3** Attach the triangular assemblies to the FEE support (Figure 9.13). The triangular assembly should lie flat against the inside groove of the FEE support. The fastener stack should be ordered as follows onto the long set screw: triangular assembly, washer, nut. To best maintain stability, let the triangular assemblies lean on the set screw hub, which is already on the mast.
- **Step 4** Attach the triangular assemblies onto the triangular supports using the socket wrench (Figure 9.14).
- **Step 5** Place the FEE on the FEE support so that the screws go through the copper holes (Figure 9.15).

Figure 9.12. Mounted FEE support.

Figure 9.13. Mounted triangular assembly. In the figure, only one of four assemblies is installed. All four are installed the same way. Do not proceed to step 4 until all four are installed.

Figure 9.14. Triangular Assembly Stabilizing. (Left) The socket wrench is used to tighten the bolts. (Right) What the assembly should look like if tightened correctly.

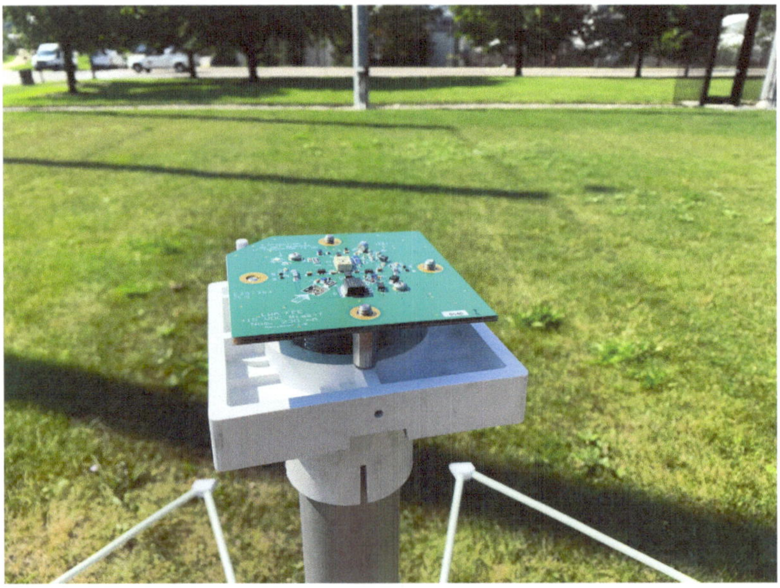

Figure 9.15. FEE mounting. To ensure you have correctly mounted the FEE, check to see if the chamfered (cut out) corner aligns with the post.

- **Step 6** Secure the FEE by inserting a washer and screwing in a nut on top (Figure 9.16).
- **Step 7** Place the FEE cover on top of the FEE support to shield the FEE. Secure the FEE cover with fasteners (Figure 9.17).

Figure 9.16. Mounted and secured FEE. The washer and nut should be the first hole directly clockwise to the chamfered corner.

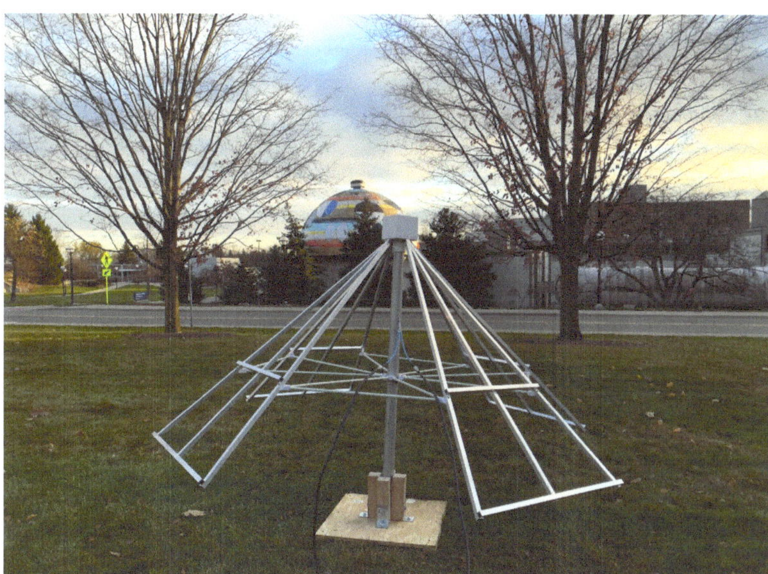

Figure 9.17. Covered FEE setup.

9.2.4.2 Power on Sequence

In order to power on the system, follow the instructions below. Refer back to Figure 9.6 to ensure you are using the right devices in the right order. Make sure that all devices are powered off before beginning.

1. Plug the power supply output connector of the 12V DC nominal power supply into the power input terminal on the Callisto Receiver.
2. Repeat step 1 for the second receiver.
3. Plug the power supply output connector of the 15V DC regulated power supply into the power input terminal on the Up-Converter.
4. Repeat step 3 for the second Up-Converter.
5. Plug the power supply output connector of the 15V DC regulated power supply into the power input terminal on the LWAPC.
6. Connect all power supplies to its AC source.
7. Switch on the power in the following order: Callisto Receiver 1, Up-Converter 1, and N–S power on the LWAPC.
8. Validate sufficient signal.
9. Switch on the power in the following order: Callisto Receiver 2, Up-Converter 2, and E–W power on the LWAPC.

9.2.4.3 Data Collection

The same Callisto Installer Guide,[2] used to set up the PCs, includes an in depth description of how to record data. Follow those steps to begin data collection. When you wish to stop data collection, follow the instructions laid forth in the document, then power off the computers, and proceed to the power off/disconnection sequence below.

9.2.4.4 Power Off/Disconnection Sequence

Similar to the power on sequence, the power off sequence must be followed in order to ensure that none of the components are damaged. Do not disconnect any cables while the power is switched on:

1. Switch off the power on the instruments in the following way: LWAPC N–S, Up-Converter 1, Receiver 1.
2. Switch off the power on the instruments in the following way: LWAPC E–W, Up-Converter 2, Receiver 2.
3. Disconnect all power supplies from its AC input.
4. Remove power supply output connector from theLWAPC power input terminal.
5. Repeat step 4 for each instrument power input in this order: Converter 1, Receiver 1, Converter 2, and Receiver 2.
6. Disconnect each coaxial cable from its input and output connection. The order does not matter.

[2] https://www.e-callisto.org/software/CallistoInstallerGuide.pdf

9.3 Student Involvement in the Project

9.3.1 Examples of Student Roles in Designing, Fabricating, Testing, and Operating the LWA Array

Students ran multiple cohorts of undergraduate and graduate students through the University of Michigan College of Engineering Multidisciplinary Design Program. The experience included software and hardware design, development, extensive testing, and deployment with the goal of exposing the students to developing high technology readiness level (TRL) systems meeting science-level data collection requirements.

The cohorts developed additional science analysis software to digitally re-point the SunRISE array at objects other than the Sun. Altogether students deployed two antenna arrays at Peach Mountain and one antenna array in Michigan's Upper Peninsula, Marquette High School, Marquette, MI.

9.4 Organization Chart

80 students have been part of the SunRISE LWA team. The team is divided into five sub-teams:

1. Mechanical team;
2. Software team;
3. System integration team;
4. Electrical team;
5. Science team.

9.5 Educational Benefits and Skill Development

Participation in ground-based radio antenna projects offers students a multifaceted learning experience that fosters technical proficiency, creativity, and teamwork. By engaging in mechanical, software, electronic, and system integration tasks, students acquire practical skills, theoretical knowledge, and professional competencies that are essential for success in STEM disciplines and beyond. These projects empower students to explore their interests, pursue lifelong learning, and make meaningful contributions to the field of radio astronomy and telecommunications.

Students involved in mechanical tasks gain hands-on experience in designing, fabricating, and assembling antenna structures. They learn fundamental engineering principles related to material selection, structural integrity, and mechanical stability. Through this aspect of the project, students develop proficiency in CAD modeling, precision machining, and mechanical prototyping, honing their skills in problem-solving, critical thinking, and project management.

The software aspect of ground-based radio antenna projects exposes students to programming languages, simulation tools, and data analysis techniques. Students acquire proficiency in software development for antenna control, signal processing, and data visualization. They learn to utilize software platforms such as MATLAB, Python, and simulation software for electromagnetic analysis. Engaging in software work enhances students' computational skills, fosters innovation in algorithm design, and cultivates a deeper understanding of signal processing concepts.

Involvement in electronic tasks enables students to explore the principles of circuit design, RF electronics, and signal transmission. They learn to design and implement RF amplifiers, filters, and frequency modulation circuits. Students gain hands-on experience in soldering, PCB layout, and component-level troubleshooting. Through electronic work, students develop practical skills in circuit prototyping, instrumentation, and measurement techniques, preparing them for careers in electronics engineering and telecommunications.

System integration tasks provide students with opportunities to integrate mechanical, software, and electronic subsystems into a cohesive antenna system. Students learn to interface hardware components with software control systems, calibrate antenna parameters, and optimize system performance. They gain insights into system-level design considerations, such as power management, thermal regulation, and environmental factors. System integration work enhances students' abilities to collaborate effectively in multidisciplinary teams, communicate technical concepts, and adapt to evolving project requirements.

9.6 Challenges and Experiences

Throughout the project there have been many challenges and last minute changes, during either equipment setup, or long term difficulties. The most common issue that is run into is the software and correctly setting the equipment up to collect data. These issues are documented and corrected so future team members can also solve the issue if it ever occurs again.

One major issue that arose was the system we were using to run the e-Callisto software. At the time we were using Raspberry Pis in order to receive the data and also upload online. During a test to collect data for about one week, the Raspberry Pi randomly stopped working two days in. The software team discovered that the SD card had become corrupted and we had to order a new one in order for it to start working again. However even with fresh SD cards, the Raspberry Pi's performance would seemingly decrease at random. That, combined with the relatively low storage capacity of the Raspberry Pis led the systems integration subteam to conduct a trade study for a replacement computer system. They concluded that the Beelink Mini S12 computer was the best suitable replacement, due to its large storage, Windows 11 processing system, and available ports. After the system was set up, the new computer was working better than the Raspberry Pi, with a more friendly interface and setup process. Having backup plans and workarounds to major stops in the project was a great learning experience for the students to have. (Note that in other radio astronomy contexts, as in Chapter 5, the Raspberry Pi is the near-ideal instrument for data collection.)

As the final enclosure for the system reached its final model, construction and installation of the AC unit also began. The precision of the drilling needed to be done was incredibly important or the heating and cooling unit would not properly fit to the enclosure. With careful marking and labeling before making cuts, the team was able to obtain an accurate location for the AC unit on the enclosure. However, some holes that were drilled were still slightly off. Many of the screws had to be tightened in sideways, leaving concern for the longevity of the mounting. Similar problems also arose with how other items were planned out and applied. The insulation material was, at first, glued on improperly which led to the heavy duty adhesive material sticking the enclosure to the ground. Additionally, the weather proofing of the enclosure was not sufficient and some water was noticed on the inside of the enclosure. After the discovery of these oversights, a new manufacturing guide was developed to avoid these mistakes again in the future. Keeping detailed documentation of all processes is something that all the students are able to learn and will be important throughout the rest of their careers.

9.7 Summary of Student Benefits of Solar Observations with LWA Antennas

Our primary responsibility involves the maintenance and enhancement of the Long Wavelength Array (LWA) antenna system. Undergraduates will learn to troubleshoot, repair, and optimize the antenna system to ensure its continued functionality for ongoing research endeavors. This hands-on experience will provide invaluable insights into antenna design, signal processing, and radio astronomy techniques.

We aim to foster a culture of active participation and contribution to space-based missions and related projects among undergraduate students. Through collaborative efforts and interdisciplinary approaches, students will have the opportunity to engage in satellite communication, CubeSat development, and other space exploration initiatives.

9.8 Testimonials and Experiences of Student Participants

9.8.1 Student #1

I have been part of the SunRISE LWA project for two years now. I first started working on doing research over thermodynamics to properly keep our hardware in our enclosure during the winter at the proper temperature. As a first year student and researcher, I learned a lot of technical and also communicational skills during our weekly meetings. It became important to properly communicate to both other team members and team leads so we all knew what was going on in the project. I have now previously worked as the mechanical team lead, and am currently the systems integration team lead as well. There have been a lot of opportunities for leadership, with preparing onboarding new members and also making important decisions on project deadlines. Overall being my first research project in space science, it has been a great introduction into the field, and also has furthered my passion for research.

9.8.2 Student #2

I was first made aware of the SunRISE LWA project as a part of the University of Michigan's College of Engineering Honors Program. During one of our seminars, we had a guest speaker come in and introduce the idea of MDPs (Multi-Disciplinary Projects). These projects included both industry and faculty research projects and were designed as a way of getting hands-on experience doing real engineering with a diverse group of students. Everyone in the Honors Program was required to complete a capstone project, as a sort of culmination of all the lessons learned and work done. We were told that anyone who participated in an MDP would be able to use that as their capstone projects. Being a sophomore with no clear project in mind, I thought I would take a look at the project options to see if anything struck my interest. Luckily for me, I ran into two of the SunRISE team leads at that point in time, who gave me the rundown on the project and everything that came with it. After being sold on the research being done, I joined the team and stayed on for three years until I graduated this past December (2023). During that time, I started as a member of the mechanical team, working on designing a new permanent structure to hold and stabilize the antenna mast. In the following years, I would take the reins as the new Mechanical Subteam lead and then, subsequently, the Systems Subteam lead. I consider my time on the team integral to the growth I experienced through college. Although classes can teach you formulas and theory, there is something about hands-on learning that resonates with me, and with lots of other students. The research and engineering being done for this project gave opportunity for students like me to experience hands-on engineering, to deal with overcoming adversity, and to improve interpersonal skills. I can say with certainty that I am a more effective communicator now than when I joined the team, and much of what is responsible for that growth is the team. Between the full team meetings, the subteam meetings, and the trips to Peach Mountain, a lot of time was spent focusing on taking a big problem and breaking it down into smaller bite-sized pieces that would then be delegated to individual students. Being a team lead, much of that breaking down and delegating was my responsibility. As a result, I learned how better to introduce topics to those previously unfamiliar, how to talk to a specific audience, and how to evaluate how someone's specialized skills would best help the team. This is not to mention all the technical skills, like CAD, machining, and setting up autonomous data collection that my time on the team has gifted me with. When I got to the end of my last semester, those technical and communication skills were able to assist me during my honors capstone presentation. Being able to present the work that I, and all of the team members had done was a full circle moment for me since I was now hoping to be the inspiration that would cause the next generation of students to join the project team. Luckily for me, and for the project as a whole, the enthusiasm for learning and for scientific development among the team members is unmatched in any educational group I've been a part of. The best part of presenting the work was seeing people's faces light up as they began to understand the science at play, the work being done at the University of Michigan, and the importance of that work in the global astronomical society. I am forever grateful to have had the opportunity and the experience to work as a part of the team, and I think research projects like this are a fantastic way of both creating more complete engineers and inspiring interest in radio astronomy for the next generation.

References

Akhavan-Tafti, M., Soni, S. L., Higgins, C., et al. 2025, E&SS, 12, e2024EA004114.

Kasper, J., Lazio, T. J. W., Romero-Wolf, A., Lux, J. P., & Neilsen, T. 2024, 2024 IEEE Aerospace Conf. (Piscataway, NJ: IEEE) 1

Nelson, S. P. 2014, Master's thesis, Colorado State University

Part IV

Innovative Methods to Use across Various Radio Astronomy Teaching Projects

AAS | IOP Astronomy

Radio Telescope Instrumentation for Teaching

Timothy Dolch

Chapter 10

Interferometry with DLITE

Joseph F. Helmboldt, Jason E. Kooi and George A. Carson

This chapter provides an opportunity to become familiar with the inner workings of one of the most widely used types of instruments in radio astronomy: the interferometer. You will learn by doing, that is, you will build a software "correlator," which is the heart of the modern interferometer, and that correlator will be tested on an operating system. Specifically, we will use one of several installations (perhaps even at your institution) of the Deployable Low-band Ionosphere and Transient Experiment (DLITE), which is a relatively simple, four-antenna interferometer (Helmboldt et al. 2021). *Estimated Cost: $60,000 USD for DLITE station; SDR work can be done in pure software, using GNURadio.*

10.1 Introduction

Interferometers rose to prominence within radio astronomy because the diffraction limit of any telescope is $\sim \lambda/D$, where λ is the wavelength of light being observed and D is the diameter of the telescope aperture. The resolution of ground-based optical telescopes are limited by atmospheric effects to around $1''$. To achieve this resolution for the 21 cm neutral hydrogen emission line would require a telescope with a diameter of over 40 km (yes, km!), which is not practical. The interferometer allows a workaround by essentially breaking up a gigantic aperture into an array of smaller apertures, which has a combined resolution of $\sim \lambda/B_{\rm max}$, where $B_{\rm max}$ is the maximum separation between points within the array, often referred to as "baselines" (hence, the B notation).

The catch is that this is essentially like carving out giant holes within the larger aperture, which makes the response of the interferometric telescope less than ideal. This can be addressed in two ways. The first is via a process sometimes referred to as Earth rotation synthesis, which relies on the fact that antennas on the surface of the Earth move relative to a cosmic radio source due to Earth's rotation (except at the poles). This allows one to gradually fill in the synthetic aperture of the interferometer over time. The second is through so-called bandwidth synthesis in which a large

doi:10.1088/2514-3433/ae1223ch10

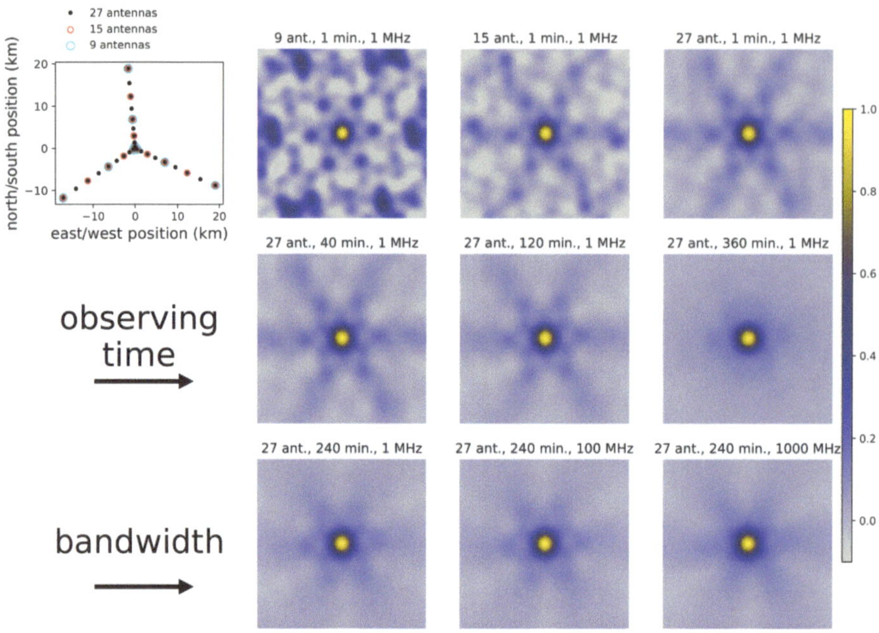

Figure 10.1. The impulse response of the VLA at 1400 MHz toward Cygnus A for different combinations of number of antennas (upper row), observing time (middle row), and bandwidth (bottom row).

fractional bandwidth is processed. For the same physical configuration of array elements, the synthetic aperture is effectively more sparse for shorter wavelengths. By processing a large bandwidth together, the aperture becomes much less sparse by including longer wavelengths while retaining the higher resolution of the shortest wavelength within the band.

This is illustrated in Figure 10.1, which shows the impulse response (i.e., the response to a point source) from the Very Large Array (VLA). A configuration at 1400 MHz toward Cygnus A for different combinations of number of antennas, observing time, and bandwidth. One can see that if enough antennas are included, the impulse response for the Y-shaped array is reasonable, even for a short integration time and relatively narrow bandwidth. As longer observing times and/or larger bandwidths are used, artifacts within the impulse response begin to disappear, and it more closely resembles the smooth response of a completely filled aperture.

10.2 How Does an Interferometer Work?

While Section 10.1 provides a top-level overview of what an interferometer can ultimately deliver, it brushes over much of the details of how such an instrument works and how the data it collects are processed. Standard texts such as Thompson et al. (1991) contain highly detailed discussions of these topics. In brief, while there are different types of interferometric telescopes that work in different ways, one of the most commonly used systems in the digital age is the so-called "FX" correlator. Most modern interferometers work by multiplying together the complex voltages from one antenna times the complex conjugate of the voltages from the second antenna of each unique baseline within the array and averaging this product in time. The FX correlator first performs a fast Fourier transform (FFT) of each signal to break it up into a preselected number of frequency channels, then multiplies the results channel-by-channel, averaging over a chosen coherent integration time (i.e., the "F" stands for FFT and the "X" indicates multiplication). These time-averaged cross-correlations are called visibilities. The primary output of an FX correlator is a dynamic spectrum of visibilities per baseline (and per polarization if multiple feeds are used per antenna).

These visibilities are, roughly speaking, the Fourier complement to the image plane, that is, a point source on the sky will look like a plane wave within visibility space that oscillates both in frequency and in time for each baseline. As a source moves on the sky, this oscillation will change in time, producing a fringe pattern that is most noticeable within the visibility phases. The exact pattern depends on the orientation and length of the baseline as well as the size(s) and sky location(s) of the radio source(s). Specifically, the visibility phase for a single source will be $= 2\pi(d\nu + f_r t) + \phi_0$ where d is the geometrical delay between the two antennas, and f_r is the fringe rate, which is $= \nu_0 \partial d/\partial t$ (ν is frequency; ν_0 is the central frequency of the band; ϕ_0 is the baseline phase at $\nu = t = 0$). The delay and fringe rate are related to source sky position by

$$d = c^{-1}(\Delta xl + \Delta ym + \Delta zn) \tag{10.1}$$

$$f_r = \lambda^{-1}(\Delta x\dot{l} + \Delta y\dot{m} + \Delta z\dot{n}) \tag{10.2}$$

$$l = \cos e \sin a = -\sin h \cos \delta \tag{10.3}$$

$$m = \cos e \cos a = -\cos h \cos \delta \sin \phi + \sin \delta \cos \phi \tag{10.4}$$

$$n = \sin e = \cos h \cos \delta \cos \phi + \sin \delta \sin \phi \tag{10.5}$$

$$\dot{l} = -\omega_e \cos h \cos \delta \tag{10.6}$$

$$\dot{m} = \omega_e \sin h \cos \delta \sin \phi \tag{10.7}$$

$$\dot{n} = -\omega_e \sin h \cos \delta \cos \phi \tag{10.8}$$

Within these equations, c is the speed of light, $\lambda = c/\nu$, Δx is the difference in east/west positions of the two antennas, Δy is the same in the north/south direction, e and a are the source's elevation and azimuth, h and δ are the source's hour angle and declination, ϕ is the latitude of the telescope, and ω_e is the angular rotation rate of the Earth. These equations also help demonstrate practical concerns regarding choices for parameter settings within the FX correlator. For instance, the largest baseline delay the system can handle will be $(2\Delta\nu)^{-1}$, where $\Delta\nu$ is the frequency channel width (i.e., bandwidth divided by number of channels). The maximum delay a source can have depends upon the size of the antenna field of view and the baseline length. If the antenna "sees" the entire sky, then $d_{max} = B/c$, where B is the baseline length. That ignores any instrumental delays resulting from, e.g., different cable lengths. If these are known, they can be corrected for within the correlator before the signals are multiplied or after the fact if the frequency channels are sufficiently narrow. Equations 10.1 – 10.8 also demonstrate that individual sources can be located with a sufficient amount of data from a single baseline. For example, by performing a two-dimensional FFT of visibilities over a selected span of time for a single baseline will yield a map of source intensity as a function of d and f_r, which can be translated to sky positions for that baseline.

10.3 DLITE as an Instructional Interferometer

While there are many interferometric radio telescopes in operation around the world, getting hands-on experience with the inner workings of these systems is hard to come by since they are typically large, complex, and expensive. To provide the opportunity to obtain such experience, we are leveraging the much simpler DLITE platform that was originally developed as a deployable system, optimized to characterize disturbances within Earth's ionosphere via their impacts on bright cosmic radio sources in the ~20–50 MHz frequency regime. As detailed by Helmboldt et al. (2021), a DLITE system consists of four inverted vee dipole antennas designed for the Long Wavelength Array (LWA; Taylor et al. 2012). There are around 6–7 sources that are bright enough for DLITE to use for ionospheric monitoring. To adequately resolve these from one another and to mitigate the impact of extended Galactic emission, the baseline lengths need to be ~200–500 m long. The optimum configuration is to have three antennas at the points of an equilateral triangle that is ~350–500 m on a side with the fourth antenna in the center. Other configurations can be accommodated as needed. As of this writing, there are four working DLITE arrays in Maryland, Florida, New Mexico, and Ohio, the latter of which was installed almost entirely by students.

In addition to having a small number of antennas/baselines, the other aspect of DLITE that makes it conducive to hands-on learning is that the receiver system consists of software defined radios (SDRs) that are controllable/programmable via the GNURadio software environment.[1] GNURadio is open source and in addition to interfaces for multiple types of SDRs, it also contains multiple processing blocks that can be tied together via an interactive flowchart using the gnuradio-companion program. After assembling and running a processing chain, a Python script is generated that can be used outside of gnuradio-companion. Within this script, each independent path within the flowchart is assigned its own thread, and multi-threading is used to maximize CPU usage. While the script is written in Python, the code behind each of the blocks is typically written in C/C++, making it much

[1] https://www.gnuradio.org

more computationally efficient than a pure Python script. Tutorials for using gnuradio-companion and for writing and incorporating one's own processing blocks can be found here:

https://wiki.gnuradio.org/index.php/Tutorials

The FX correlator currently used by DLITE systems was written entirely in GNURadio with some modification of the Python script after the fact to enable the use of a configuration file for inputs as well as a specified observation duration. The correlator outputs visibilities for each of the six baselines in the array for a specified bandwidth (usually 8.333 MHz), number of frequency channels (usually 512), and integration time (usually 1.024 s). This is nominally done separately for the north/south (X) and east/west (Y) linear polarization feeds, but cross products between polarizations can be generated as well (e.g., X-polarization from one antenna times Y-polarization from the other). The visibilities are written to files as 32-bit, unformatted binary numbers with the real and imaginary parts interleaved.

An example of the correlator output from the DLITE system in New Mexico is shown in Figure 10.2 for a single baseline that is 320 m long and mostly oriented north/south (antennas 2 and 4). Both the amplitudes and phases of the visibilities are displayed as images as functions of frequency and time for a one-day collection (a full sidereal day; not quite 24 hours). One can see interesting fringe patterns throughout the observation resulting from multiple sources rising and setting through the antennas' fields of view. The combined frequency response of the antennas and low-pass filters used by the system (cutoff frequency of 50 MHz) is also apparent as are several instances of narrow- and broad-band interference. To illustrate the detection and identification of sources using the delay/fringe-rate method described in Section 10.2, a two-dimensional FFT was applied to the visibilities from the first hour of the observation. The resulting image is shown to

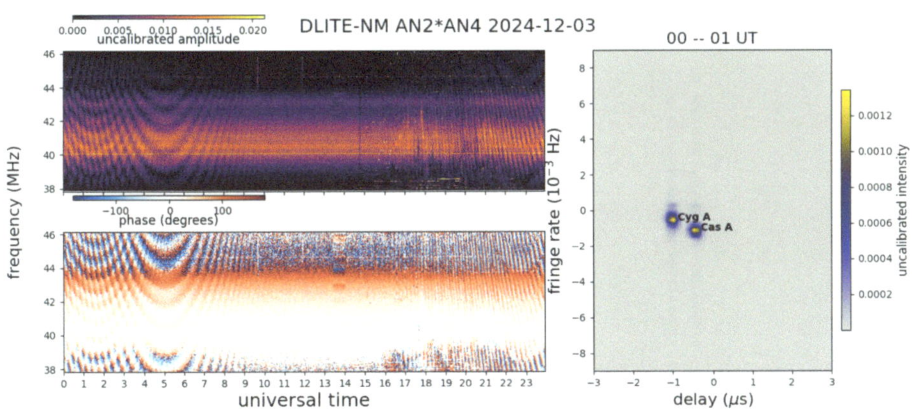

Figure 10.2. Example outputs from the New Mexico DLITE correlator for a one-day observation. Four two antennas forming 320 m long, mostly north/south baseline (antennas 2 and 4), the visibility amplitude (upper left) and phase (lower left) as functions of time and frequency. A fringe-rate/delay image generated with the first hour of data is shown to the right where the bright sources Cygnus A and Cassiopeia A are apparent and distinguishable.

the right with the expected fringe rates and delays (including a differential cable delay of $-0.9\mu s$) for the bright sources Cygnus A and Cassiopeia A (Cyg A and Cas A, respectively) labeled.

10.4 Build Your Own Correlator

Now that you have been given a crash course in interferometry generally and DLITE specifically, it is time to learn more by doing. Using the information provided here and the gnuradio-companion software, you will build your own FX correlator and have it tested with an operational DLITE system. First, to get you started, a GNURadio companion file, dlite_rxonly.grc, has been provided that has a source block added for the SDRs. Getting the settings correct for this block is a bit tricky and not particularly relevant to learning about correlator signal processing. A screenshot of the contents of the file within the gnuradio-companion interface is shown in Figure 10.3 with some helpful notations.

All of the processing blocks you need to complete the necessary steps in the correlator processing chain (refer back to Section 10.2) are available and can be attached to one another to complete the flowchart. Pay close attention to the types of inputs/outputs required for each processing block, e.g., some may require data streams while others may require arrays of numbers. Some conversion between types may be needed. Also keep in mind that the FFT implementation used by GNURadio requires array lengths that are powers of two (i.e., 2, 4, 8, 16, 32, 64, etc.). The end

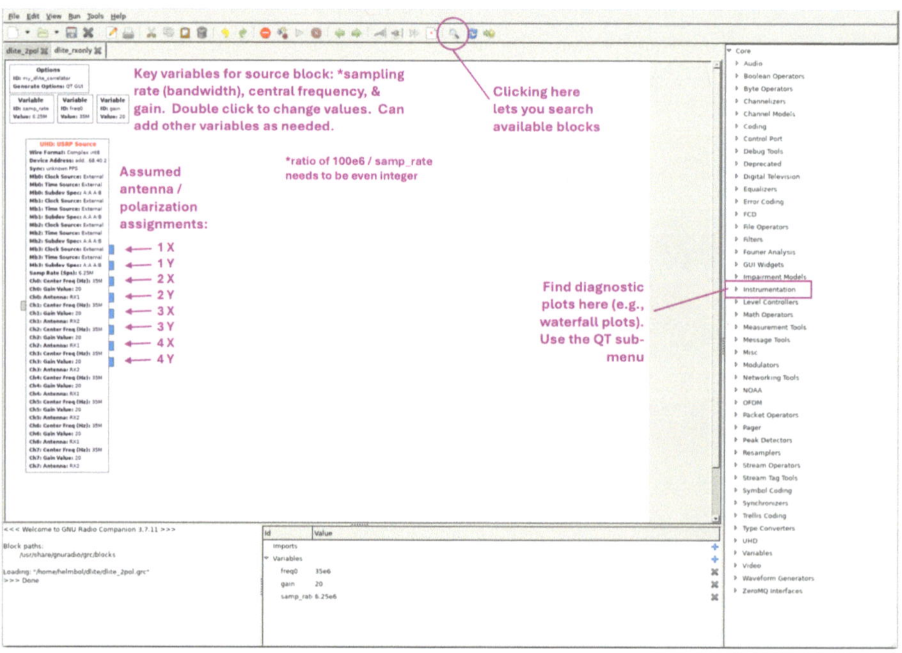

Figure 10.3. Screenshot of the contents of the dlite_rxonly.grc file within the gnuradio-companion interface with helpful notations.

products should be binary output files, one for each baseline and polarization product. Note that each of the four SDRs has two inputs, one for each polarization feed of each antenna. The SDR source block treats these as a single, eight-output receiver system. As indicated in Figure 10.3, from top to bottom, these eight outputs are assumed to be the X-polarization from antenna 1, the Y-polarization from antenna 1, the X-polarization from antenna 2, and so on. You can choose to build a correlator with either two (XX and YY) or four (XX, YY, XY, and YX) polarization products. Note that one typically has to sacrifice bandwidth to have all four polarization products due to CPU limitations.

Once you have completed the flowchart, your new correlator can be tested on an active DLITE system by submitting it to mailto:dlite_correlator_testing@us.navy.mildlite_correlator_testing@us.navy.mil. You can send either the gnuradio-companion file (.grc extension) or the Python file it generates (.py extension). If you submit the latter, it should be sent as a text file (i.e., change the file extension to .txt). Instructors can check constructed correlators prior to submission by comparing the GNURadio flowchart to the example shown in Figure 10.4. If the test is successful, and you feel adventurous, you may consider trying to write and implement your own custom processing blocks to help make the correlator more efficient and/or add functionality. This should be done in C/C++ to maximize computational efficiency. As mentioned previously, a helpful and extensive tutorial for this process can be found at:

https://wiki.gnuradio.org/index.php/Tutorials.

One type of block that might be tried is a simple operation to replace with zeros any voltages with amplitudes above a set threshold as a way to mitigate strong,

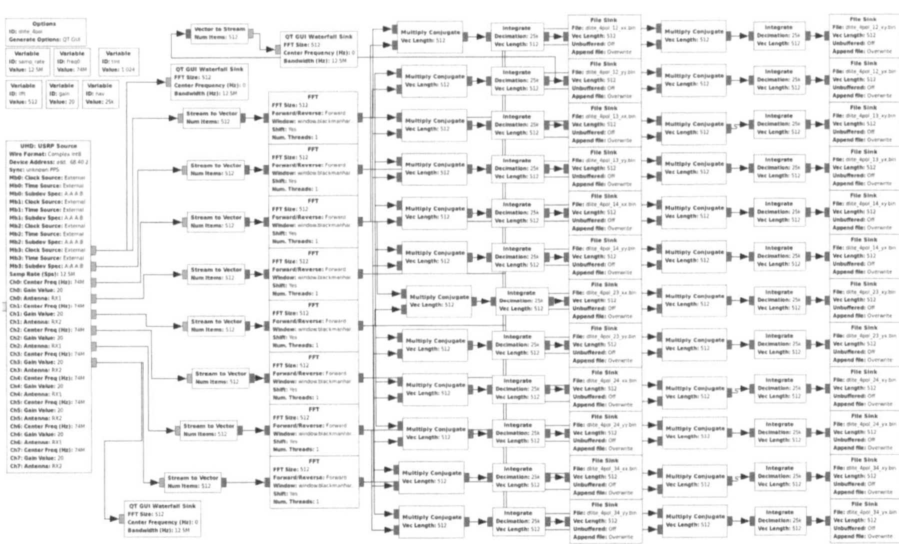

Figure 10.4. Screenshot of a flowchart created within gnuradio-companion for a working correlator with all four polarization products and a real-time diagnostic plotting window.

short-duration interference. One might also consider combining several of the processing steps into a single block to simplify the flowchart.

Additionally, Young et al. (2025) contains a linked series of Jupyter notebooks with interferometry exercises based on DLITE data that will provide a helpful supplement to the material in this chapter.

References

Helmboldt, J. F., Markowski, B. B., Bonanno, D. J., et al. 2021, RaSc, 56, e2021RS007298

Taylor, G. B., Ellingson, S. W., Kassim, N. E., et al. 2012, JAI, 1, 1250004

Thompson, A. R., Moran, J. M., & Swenson, G. W. 1991, Interferometry and Synthesis in Radio Astronomy (Malabar, FL: Krieger)

Young, O. R., Dolch, T. E., Helmboldt, J. F., et al. 2025, ApJ, 979, 203

Chapter 11

The Low-Frequency All-Sky Monitor

Timothy E. Dolch, Louis P. Dartez and Joseph F. Helmboldt

This chapter introduces a cost-effective way for students to get hands-on experience by building a low-frequency radio telescope. Implementation of this project will expose learners to a range of topics including hardware and electronics, basic radio signal processing, and radio interferometry. In the following sections, we will introduce the reader to the Low Frequency All Sky Monitor, or LoFASM (Dartez 2014; Handzo 2015), a distributed radio array optimized for ionospheric and transient spectral studies. The experimental goal is to observe the brightest so-called "A-team" sources: the radio galaxy Cygnus A and the supernova remnant Cassiopeia A. *Estimated Cost: $11,200 USD.*

11.1 Introduction

The LoFASM system is intentionally engineered to minimize reliance on costly infrastructure such as GPU clusters, high-speed communication networks, and large-scale digital processing hardware, making it cost-effective and easily deployable. This deliberate avoidance of costly components, makes LoFASM particularly well-suited for deployment at educational institutions. Its accessible design allows students to gain hands-on experience with radio astronomy hardware, signal processing, and observational techniques without the need for high-overhead infrastructure. For a breakdown of components and their costs, see Table 11.1.

In Section 11.2 we provide background on the design and implementation of the LoFASM system as a radio telescope. In Section 11.4 we propose a pared-down version of LoFASM that can be implemented at an institution by student-groups to explore what it takes to build a functioning radio telescope, record data, and process that data to identify transient signatures (e.g., ionospheric activity). The Red Pitaya exercises were developed in the context of the course "Signal Processing in Astrophysics" in the Hillsdale College Physics Department.

doi:10.1088/2514-3433/ae1223ch11

Table 11.1. Parts Used to Construct LoFASM-Lite, in USD

Item	Number Used	Manufacturer	Current Price	Notes
Inverted vee dipole antennas	7	Burns Industries	US $495.00 per antenna	Antennas used for LWA, includes mast
Front end electronics (FEEs)	7	Burns Industries	US $230.00 per unit	
Ground stake	7	Burns Industries	US $41.00 per unit	All antenna elements (dipole antenna, FEE, and ground stake) can also be purchased as a kit from Reeve Radio Observatories
Times Microwave Systems Inc. LMR-240DB Flex Cable	8	TESSCO Technologies Incorporated	US $1.07 per foot	Eight required per polarization
Times Microwave Systems Inc. SMA Male for LMR-240DB Flex Cable	8	TESSCO Technologies Incorporated	US $8.47 per unit	Eight required per polarization
Times Microwave Systems Inc. N Male for LMR-240DB Flex Cable	8	TESSCO Technologies Incorporated	US $10.79 per unit	Eight required per polarization
Bias Tee	2	Mini-Circuits	US $89.95 per unit	Part number: Z3BT-2R15G+; Two required per polarization
Amplifier	2	Mini-Circuits	US $111.95 per unit	Part number: GALI-74; Two required per polarization
High-Pass Filter	2	Mini-Circuits	US $110.90 per unit	Part number: SCHF-17+; Two required per polarization
Low-Pass Filter	2	Mini-Circuits	US $102.60 per unit	Part number: SCLF-65+; Two required per polarization
SMA Cables	6	Mouser	US $21.17 per unit	Part number: 523-135101-04-M0.25; Six required per polarization
Red Pitaya 125-14	1	Red Pitaya	US $490.00 per unit	Research was on the now discontinued 125-10, but the 125-14 is the closest currently available

Junction Boxes	7	Burns Industries	US $12.00 per unit	
DC Power Supply	1	SRA Soldering Products	US $259.99 per unit	Model KD3303P
Rack mounted computer	1	SuperLogics	US $2,494 per unit	Model SL-2U-MH510I-GD
Lightning Arrestor	2	Tessco	US $43.78 per unit	0*–7 GHz Lighting Arrestor with N Female - N Female Bulkhead connectors; 2 required per polarization
Ground Screen	1	Everbilt	US $156.98 per unit	5 ft. × 100 ft. 14-Gauge Galvanized Steel Welded Wire
Wire Rope Clips	28		US $5.29 per unit	
Tent Stakes	28		US $1.58 per unit	
Enclosure	1	Nema Enclosures	US $343.28 per unit	
Combiner Box	1	Pulsar Microwave	US $425.00 per unit	Part Number: P6-A48-434/1N; 1 required per polarization

11.2 Background on the Low Frequency All Sky Monitor

The Low Frequency All Sky Monitor (LoFASM) is a distributed radio telescope array designed to observe low-frequency transients in the 10–88 MHz band (Dartez 2021). It uses dual-polarization dipole antennas similar to those deployed in the Long Wavelength Array (LWA), taking advantage of their broad frequency coverage and stable beam patterns. LoFASM stations are equipped with digital backends based on field-programmable gate arrays (FPGAs) capable of real-time spectral analysis and data recording with high time and frequency resolution.

11.2.1 Antenna Configuration

Each LoFASM station consists of 12 cross-dipole antennas arranged in a pair of concentric rings (see Figure 11.1). The antenna signals are digitized and processed locally, and the data products are stored for offline analysis. Coincidence detection across stations enables the rejection of local radio frequency interference and supports basic interferometric techniques such as delay and fringe analysis.

The LoFASM array has a single zenith-pointing beam. In contrast to most radio arrays, LoFASM's antenna signals are combined *in analog*. This design choice, while making it impossible to electronically steer the main beam, was key to minimizing equipment cost.

The electric field incident on the jth antenna due to an incident plane wave is

$$\vec{E_j}(\omega, t) = E_0 \hat{p} e^{i(\vec{k} \cdot \vec{x_j} - \omega t)}, \tag{11.1}$$

where E_0 is the electric field amplitude, $\hat{p}$ is a unit vector that points in the direction of the field polarization, $\vec{k}$ is the wave vector pointing in the wave's propagation,

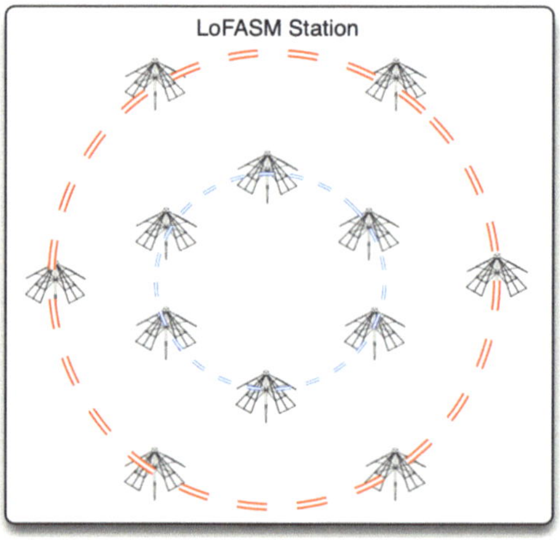

Figure 11.1. The antenna configuration of a LoFASM station. Each station consists of 12 LWA-style active cross-dipole antennas. The antennas are arranged as a pair of concentric rings, each with six elements. Reproduced with permission from Dartez (2014).

and ω is the angular frequency. The voltage induced at the terminals of each antenna element is

$$V_j(t) = G_n(\vec{k}, \vec{p})E_0 e^{i(\vec{k}\cdot\vec{x_j} - \omega t)}, \tag{11.2}$$

where $V_j(t)$ is the time-dependent voltage, $G_n(\vec{k}, \vec{p})$ is the antenna gain, and n denotes the polarization of the antenna element.

The power pattern of the coherently summed signals from all antennas of the same polarization is given by:

$$P_n \sim \sum_{i,j} V_i V_j^* = |G_n|^2 |E_0|^2 \sum_{j,l} e^{i\vec{k}\cdot(\vec{x_j} - \vec{x_l})}, \tag{11.3}$$

where l and j are antenna indices and the G_n is assumed to be identical for all elements in the array. Assuming an angular antenna beam pattern that is proportional to $\frac{1+\cos^2\theta}{2}$, the normalized beam pattern of the LoFASM array is

$$B(\theta, \phi) = \left(\frac{1 + \cos^2\theta}{2}\right)\frac{1}{N^2}\sum_{j,l} e^{i\vec{k}\cdot(\vec{x_j} - \vec{x_l})}. \tag{11.4}$$

To minimize contributions from local RF originating from the horizon and to maximize sensitivity to RF signals originating from the sky, the LoFASM configuration was chosen to minimize the fractional beam pattern sensitivity to signals at low altitude. The target beam pattern used is H/Ω_0, where $\Omega_0 = \int B\mathrm{d}\Omega$ is the solid angle subtended by the LoFASM beam pattern and H is

$$H = \int_{\phi_0}^{\pi/2}\int_{\theta=0}^{\pi} B(\theta, \phi)\sin\theta\mathrm{d}\theta\mathrm{d}\phi \tag{11.5}$$

where ϕ_0 is a zenith angle that represents the adjusted horizon (Miller 2012). The radii of the two LoFASM rings were chosen to maximize RF rejection at 20 MHz.

11.2.2 Geographic Placement of Stations

LoFASM comprises five geographically distributed stations: Port Mansfield, Texas, Hillsdale College in Hillsdale, Michigan (Figure 11.2), the North Arm of Very Large Array in New Mexico, the Green Bank Observatory in West Virginia, and NASA's Goldstone Deep Space Communication Complex in California. (The NM station has been incorporated into the 64-antenna LWA-Swarm North Arm mini-station (Taylor et al. 2025). LoFASM's low cost enables 13-antenna stations as "seeds" of eventual larger stations when more funding becomes available.) Figure 11.3 shows a map of the five stations across the continental United States. While the stations are located several thousand kilometers apart, they have overlapping views of the sky. The geographic separation enables the rejection of local noise and artifacts through anti-coincident detection, where signals appearing at only one station are treated as local interference rather than genuine signals from the sky.

Figure 11.2. LoFASM V at Hillsdale College in Hillsdale, MI (picture provided by T. Dolch).

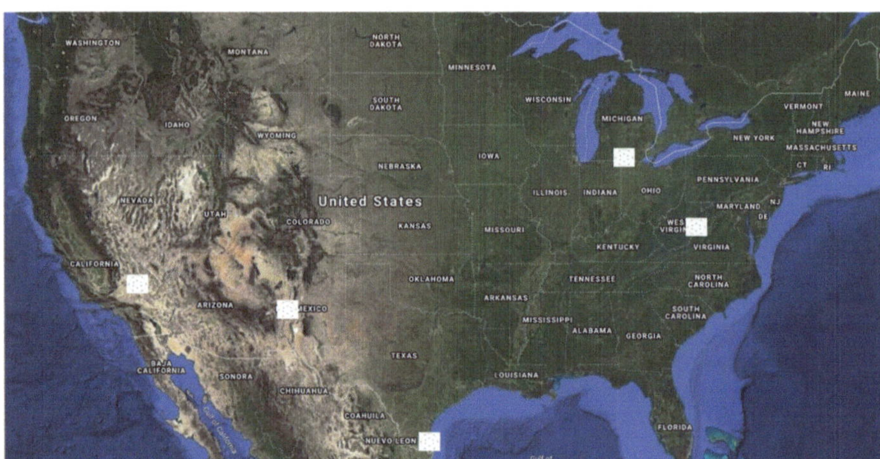

Figure 11.3. A map showing the locations of the five LoFASM stations across the continental United States. Map data © 2025 Google.

11.2.3 Relationship to Other Distributed Low-frequency Arrays

LoFASM occupies a unique niche among radio observatories. While many instruments capable of observing in the 10–88 MHz range exist (LWA, DLITE, and SunRISE have all been represented in this book), few are optimized for continuous, wideband surveillance across this band. Instead, most radio facilities prioritize higher sensitivity over narrower bands, often focusing on specific frequencies for targeted science goals. In contrast, LoFASM is deliberately engineered for ease of deployment and operation, enabling uninterrupted broadband capture across the full sub-100 MHz spectrum accessible from the ground. This design choice makes LoFASM particularly valuable for monitoring transient events, capturing rare phenomena, and serving as a persistent background observer in an underutilized region of the radio spectrum.

11.2.4 Analog Signal Chain

The LoFASM signal chain design favors analog components where feasible to minimize reliance on costly digital processing. Rather than implementing dynamic, multi-directional digital beamforming, LoFASM employs a fixed analog beam pointing at zenith. The radio frequency (RF) signals originate at the antennas. LoFASM makes use of the readily-available LWA antenna (Hicks et al. 2012). The antennas house front-end electronics boards (FEEs), which are directly connected to the cross-dipole antenna blades, that apply initial amplification to the raw RF signals incident on the antennas. From the FEEs, the signals from each ring of antennas are routed to a 6-to-1 DC-passing RF combiner via LMR-195 coaxial cable. The result is a pair of signals per ring (one per polarization). In total, four signals from the two rings are transmitted along LMR-400 coaxial cables to a control room.

In the control room, the signals are sent to the same analog signal receivers (ARX). The LoFASM ARX design, inspired by more complex LWA ARXs, involves multiple amplification and filtering stages. The first element in the ARX is a bias-T. The bias-T is a passive circuit that allows both DC power and RF signals to coexist on the same coaxial cable without interfering with each other. This is particularly useful for low-frequency radio telescopes where it is desirable to power an active antenna remotely without requiring a separate power cable. Further ARX stages are typically an alternating series of amplification and pass-band filtering to prepare the RF signals for digitization. Figure 11.4 shows an example of what the pass-band at the output of the LoFASM ARX signal chain typically looks like. See Figure 11.5 for a schematic depicting the LoFASM ARX topology.

11.2.5 Digital Backend

The LoFASM digital backend comprises of an FPGA-based processing unit and high-throughput control computer. The FPGA processing unit is a Reconfigurable Open Architecture Computing Hardware 1 (ROACH-1) board that implements a 4-input full Stokes 100 MHz correlating spectrometer. The firmware logic for the spectrometer was designed using tools provided by The Collaboration for Astronomy Signal Processing and Electronics Research (CASPER). The ROACH-1 board interfaces with dual 14-bit ADCs sampling at 200 Msps to digitize the incoming RF signals from each polarization channel. Once digitized, the data are fed into a real-time correlating spectrometer implemented with the FPGA on the ROACH-1. This spectrometer computes auto- and cross-spectra between the four input channels, producing spectral products with a time resolution of $\sim$83 ms. These process spectra are then streamed over 10 Gbps Ethernet to a host computer, where they are archived

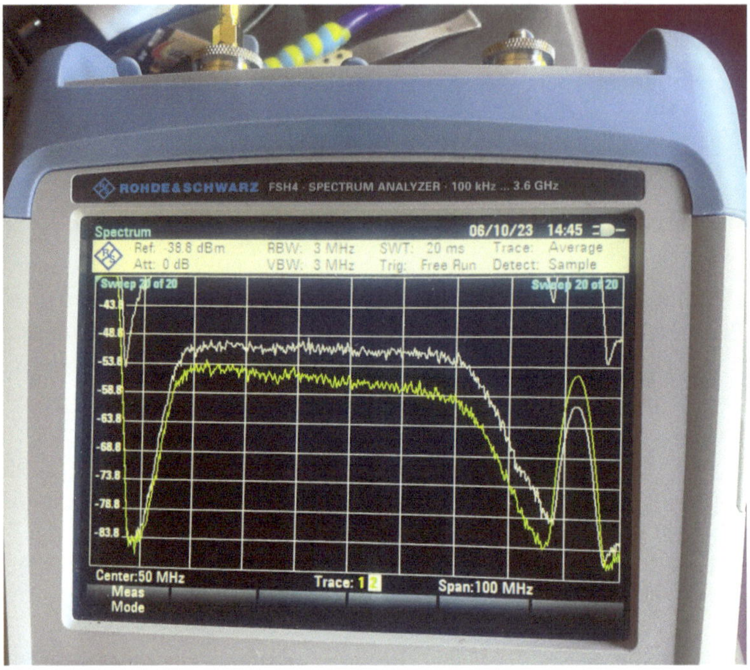

Figure 11.4. Image of a spectrum analyzer (*Rohde & Schwarz FSH4*) connected to the output of a LoFASM ARX chain. The yellow and white curves are that of a typical pass-band at the RF OUT port of a LoFASM ARX stage. The ARX implements low-pass and high-pass filtering to suppress signals outside of the 10–88 MHz window. It also provides ~26 dB of amplification.

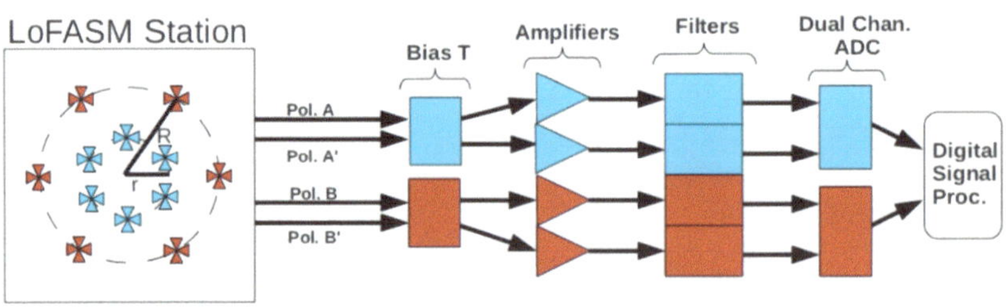

Figure 11.5. A schematic of the LoFASM signal path from the active antenna elements on the left to the digital signal processing (DSP) unit on the right. Since each element consists of a cross-dipole antenna, two linearly polarized signals are provided by each element, each with its own cabling. The signals from each polarization in each ring are combined in phase via a set of analog combiners located at the center of the array. The polarization pairs (A, A'), and (B, B') depicted in the figure represent the two resultant polarized signals from each ring (i.e., channels A and A' contain the orthogonal linearly polarized signals from the antennas in one ring, and B and B' contain the other). Once inside the electronics shelter the signals go through the "bias-T" and a few stages of filtering and amplification before being sent to the digital processing backend to be digitized.

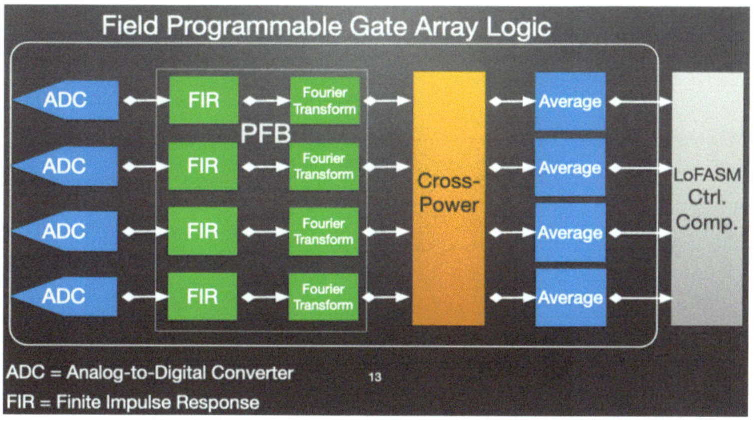

Figure 11.6. A flowchart representing the flow of operations that the LoFASM correlator firmware applies to incoming RF signals. First the signal is digitized then it is processed through a polyphase filterbank, which is an efficient digital method for slicing a wideband signal into several clean, evenly-spaced narrow-band channels. The resultant PSD is used to calculate all possible auto and cross-correlation pairs before being averaged for 83.4 ms. Reproduced with permission from Price (2018).

to disk and subjected to post-processing stages. A schematic depicting the logic that the spectrometer is based on can be seen in Figure 11.6.

For each pair of antenna inputs, $V_1(t)$ and $V_2(t)$, the correlator computes their cross-correlation

$$r(\tau) = \lim_{T\to\infty} \frac{1}{2T} \int_{-T}^{T} V_1(t) V_2^*(t - \tau)\mathrm{d}t, \tag{11.6}$$

where τ is a time delay and $2T$ is the total integration time. The cross-correlation is a comparison of signals recorded by two antennas at various time delays to identify the delay at which they are most similar. The time delay for which the two signals best match is related to the source's direction on the sky. The LoFASM backend uses ADCs that produce real-valued digital voltage samples. For real-valued signals, Equation (11.6) becomes

$$r(\tau) = \langle V_1(t) V_2(t - \tau)\rangle. \tag{11.7}$$

$$S(\nu) = \int_{-\infty}^{\infty} r(\tau)e^{-i2\pi\nu\tau}\mathrm{d}\tau \tag{11.8}$$

At a single given frequency, ν, Equation (11.7) reduces to $\frac{A_1A_2}{2}\cos(2\pi\nu\tau)$, where A_1 and A_2 are the amplitudes of the two inputs. The cross-correlation is then dependent on the delay, τ, between the two signals. In radio astronomy it is common to compute the cross-power spectral density in the frequency domain (Equation (11.8)). This is often referred to as the *visibility* in radio astronomy. Notably, $S(\nu)$ and $r(\tau)$ form a Fourier pair (per the Wiener–Khinchin theorem). Figure 11.7 shows the cross-power spectral density taken at LoFASM Green Bank over a 24 hour period.

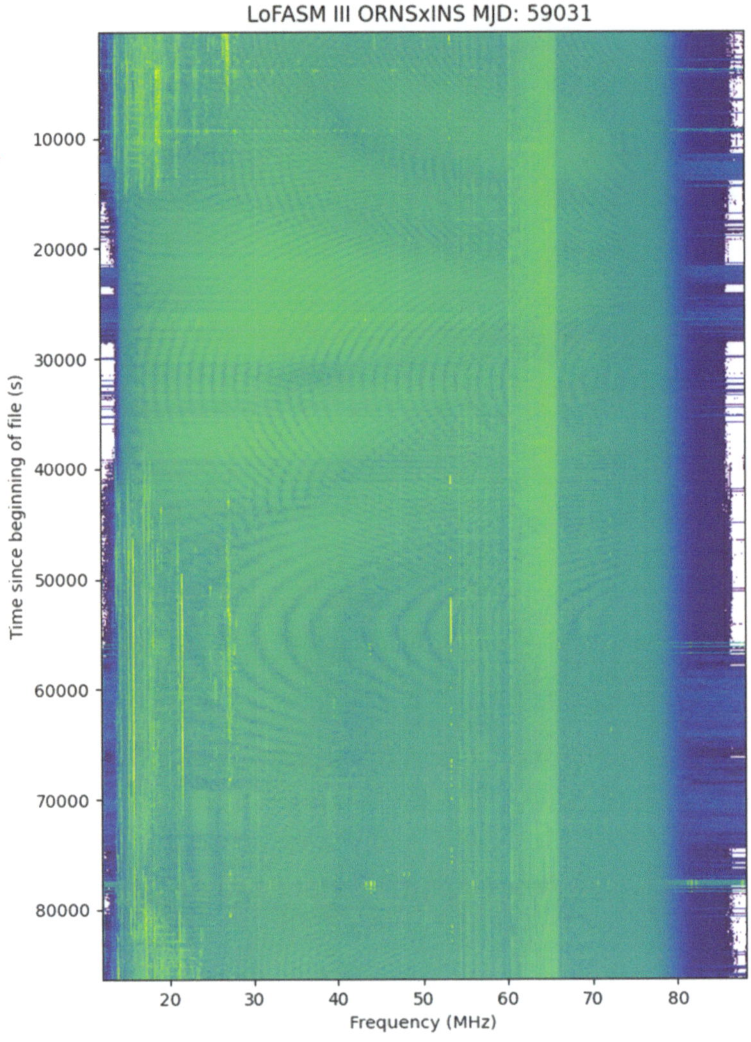

Figure 11.7. Spectrogram showing the power spectral density taken by computing the cross-correlation of the inner and outer LoFASM rings over a 24 hour period at LoFASM Green Bank. The vertical axis is time elapsed in seconds from the start of the observation and the horizontal axis is frequency in MHz. The intensity is the power spectral density in arbitrary units. The spectrogram shows some vertical features that represent radio frequency interference that is transient at low frequencies and persistent near 60–65 MHz. Also visible is an interference pattern that arises because the cross-power spectral density inputs are unique elements. The fact that the interference pattern evolves over time suggests that the source is moving with respect to the telescope elements. For a discussion on the use of LoFASM as an interferometer see Section 11.3.

11.2.6 Local Clock Sources

Key to the operation of any long-running radio frequency instrument is the availability of stable clock signal sources. LoFASM makes use of at least two independent clock signals at each installation. These are a 1 pulse-per-second (1 PPS) and a 200 Msps clock source. The 1 PPS is connected directly to the

ROACH-1 board and provides a consistent rising edge for the FPGA firmware to ensure that the FPGA logic runs synchronously and deterministically. The other clock source drives the ADCs, which are deployed on daughter cards connected directly to the ROACH-1 board.

To drive the ADC sampling LoFASM uses a commercial off-the-shelf Valon Synthesizer 5008.[1] As of this writing, the 5008 is now discontinued and has been replaced by the newer 5009 model.[2] The ADC cards on the ROACH-1 board are driven by the Valon as an external clock reference. The external and independent clock source ensures that both ADCs boards are sampling their respective inputs synchronously. In the absence of an external clock source, the ADCs are not guaranteed to sample and digitize the voltages at the RF inputs simultaneously. Minute temporal delays in the ADC voltage sampling can introduce additional jitter and phase noise into the recorded signals. The Valon synthesizers are equipped with two RF outputs. In practice each one is connected to one of the two ADC cards installed on the ROACH-1 board. The Valon models also have an RF input for an external 10 MHz (on the 5008 unit) clock reference. The 10 MHz helps keep the two RF out clock ports locked to each other, further improving the synchronization between the clock signals driving the ADCs.

11.3 Use of Outrigger for Interferometry

Interferometry is a technique that allows for multiple independent antennas separated by some distance to be used together as a single instrument by electronically (either in analog or digitally) steering a synthesized aperture. This is done by correcting for the temporal delays associated with the fact that any wave incident on the antennas in the telescope will exhibit a different time of arrival unique to each element and depends on the direction of the source of emission. The only time that the time of arrival of a signal will be the same for all elements in a group of antennas forming two-dimensional baselines is when the source is directly overhead (or somehow directly below, a situation that is mathematically possible but practically unlikely). In the case of an interferometer that consists only of two antennas separated by some fixed distance, and thus forming a single one dimensional baseline between them, the directions with equivalent delay form a cone with a specified opening angle and the baseline as its axis. For a simple schematic depicting a two-element interferometer, see Figure 11.8.

The placement of an outrigger at each of the LoFASM stations allows for interferometry to be performed by treating the main grouping of antennas as a single element and the lone outrigger antenna as the other in a two-element interferometer. Since the array-outrigger systems are oriented (mostly) East–West, we are able to easily detect stronger sources and localize them more accurately in Right Ascension than in Declination. As bright point sources rise above the horizon toward the east

[1] https://www.valonrf.com/frequency-synthesizer-4400mhz.html
[2] https://www.valonrf.com/frequency-synthesizer-6ghz.html

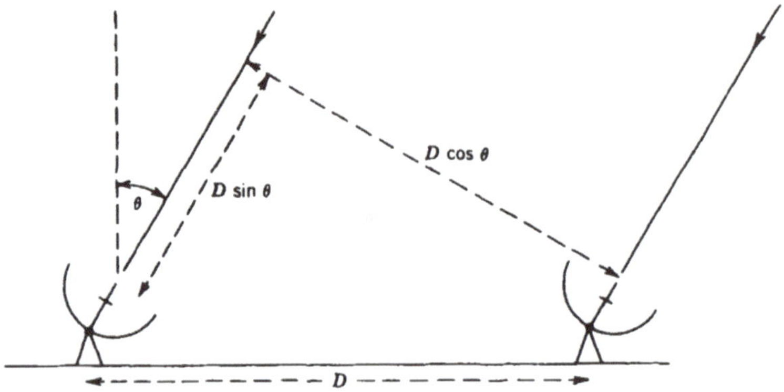

Figure 11.8. Geometry of an elementary interferometer. D is the interferometer baseline and θ is the angle between zenith and the direction of the source. Reproduced with permission from Thompson et al. (2017). © 2017 The Author(s) CC BY-NC 4.0

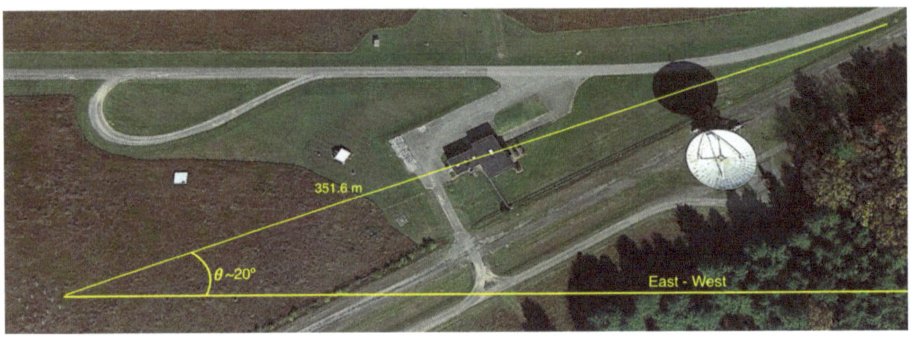

Figure 11.9. LoFASM at Green Bank depicting the main 12-element array (bottom left) and an outrigger antenna (top right). In its two-element interferometer mode, LoFASM operates with the main array as a single element and outrigger antenna as the second. The baseline separation between the outrigger and the main array is ~350 m. Map data © 2025 Google.

and move overhead the voltage as a function of time in the interferometer response, which is obtained from the multiplication of the outputs of the main array and the outrigger, will contain a corresponding interference pattern that manifests in the form of ripples in the integrated cross-power data products recorded by the LoFASM Spectrometer. Figure 11.9 shows an image of the baseline at the Green Bank LoFASM main array and its outrigger antenna.

The spacing between the oscillations of the "ripples" in these interference patterns will vary as a function of geometric delay, which advances as the source passes overhead. The geometric time delay, τ_g, which is proportional to the projection of the baseline vector onto the source direction vector, is

$$\tau_g = \frac{\hat{s} \cdot \vec{B}}{c}, \tag{11.9}$$

where $\hat{s}$ is a unit vector pointing in the direction of the source, and $\vec{B}$ is the baseline between the pair of antennas. The delay is at a maximum when the source is rising and setting and at a minimum when its altitude is at a maximum. In terms of the angle between the baseline and source direction, θ, Equation (11.9) becomes $\tau_g = B\sin(\theta)/c$. The geometric time delay, τ_g, produces an interference pattern in the visibility function with a frequency-dependent fringe phase $\varphi(t, \nu) = 2\pi\nu\tau$, which implicitly depends on the direction of the source on the sky:

$$\varphi(t, \nu) = \frac{2\pi\nu}{c} B \sin\theta(t), \tag{11.10}$$

where ν is the observation frequency in Hertz and $\theta(t)$ is the time-dependent angle between the baseline vector and the direction of the source on the sky. Figure 11.10 shows the fringes from Cygnus A in data recorded at LoFASM Green Bank (a.k.a. LoFASM III). Given a source at direction $\hat{s}$, one can calculate the geometric time delay, τ_g, by evaluating the projection of the source unit vector onto the baseline $\hat{s} \cdot \vec{B}$:

$$\tau_g = \frac{B}{c}[\cos\rho(\sin\delta\cos\phi - \cos\delta\sin\phi\cos H) + \sin\rho\cos\delta\sin H], \tag{11.11}$$

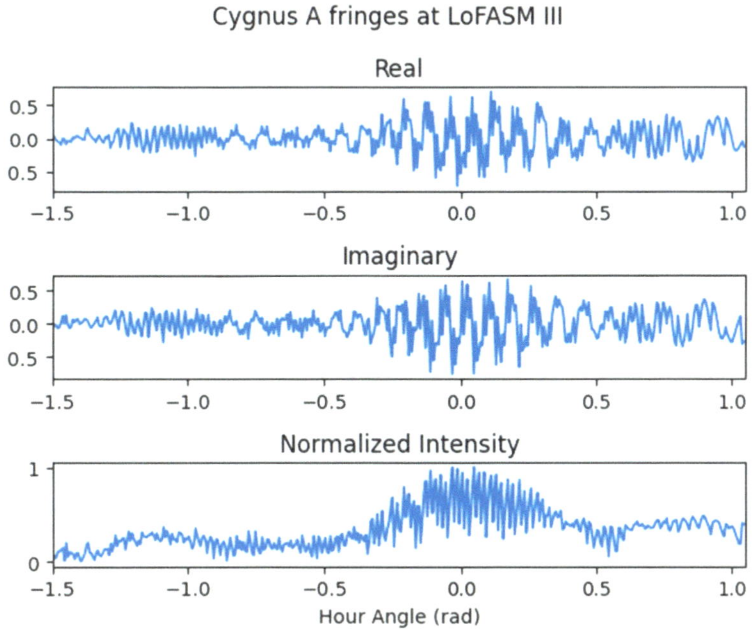

Figure 11.10. Cygnus A fringes at LoFASM Green Bank (a.k.a. LoFASM III) taken while the instrument was in its interferometer mode. The vertical axes show the real (top), imaginary (center), and normalized (bottom) intensity in arbitrary units. The horizontal axis is hour angle in radians. From Dartez (2021).

where ϕ is the observatory latitude, ρ is the angular offset between true East–West and baseline vector $\vec{B}$, H is the local hour angle, and δ is the source declination coordinate. Figure 11.11 shows a drift scan delay spectrogram taken at LoFASM Green Bank while Cygnus A and Cassiopeia A rise and set. Note, however, that that these data were taken from the low-RFI environment of Green Bank Observatory's location in the National Radio Quiet Zone.

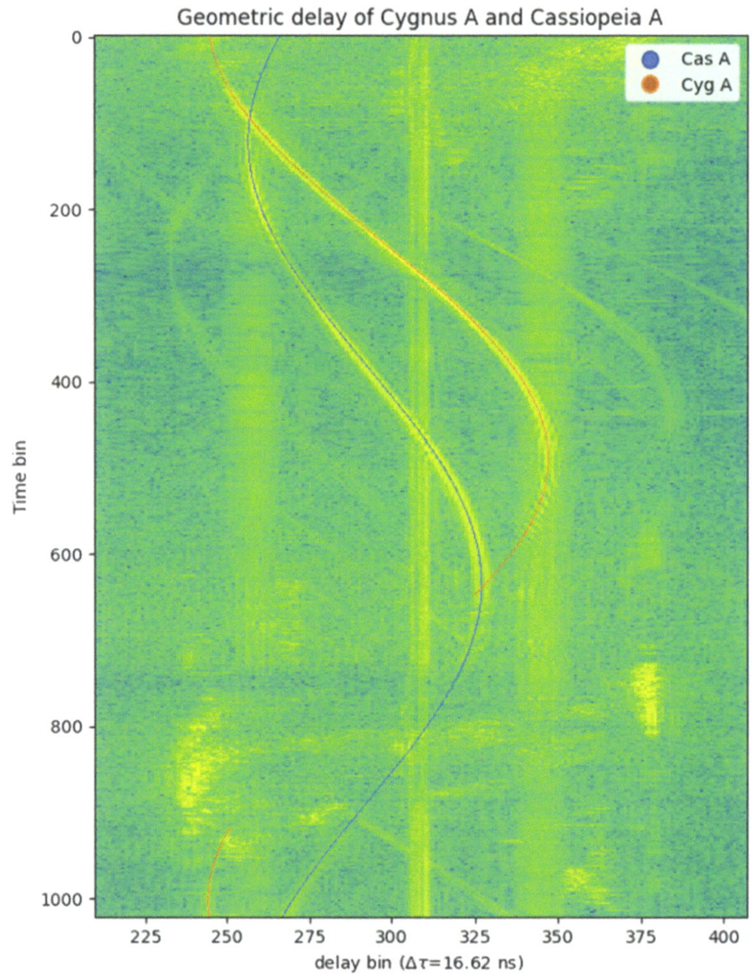

Figure 11.11. Delay spectrogram depicting the delay tracks from the two brightest sources, Cygnus A (red) and Cassiopeia A (blue). The calculated geometric delay has been calculated for both sources and overlaid on the spectrogram. The vertical axis represents elapsed time in units of sample index. The horizontal axis represents geometric delay in units of bin index. Reproduced with permission from Dartez (2021). CC BY-NC-SA 4.0

11.4 Background on the Red Pitaya

The STEMLab Red Pitaya is an accessible FPGA-based system being increasingly used in small experimental setups (Pultinevicius et al. 2023). In the context of the Low-Frequency All-Sky Monitor, this provides a low-cost backend receiver system as an alternative to the ROACH-1 board used in the original telescope design. The Red Pitaya is a signal acquisition board similar in concept to a Raspberry Pi, including a Linux operating system, except that it features a built-in FPGA. In our demonstration, we specifically use the STEMLab 125-10. The 125-10 features two input and two output channels; here we use the two input channels to connect to two widely separated antenna elements. As of the time of this writing the STEMLab 125-10 has been discontinued, but has been replaced with comparable higher-quality models such as the STEMLab 125-14. A disadvantage of the STEMLab 125-10 compared to the ROACH-11 is that its bandwidth as a spectrum analyzer is 0–60 MHz, which means the typical 10–88 MHz LWA bandwidth is not entirely covered. However, we will demonstrate that a simple interferometer experiment is possible even with a smaller bandwidth. As the cost of advanced signal processing and data acquisition systems continues to decrease, so has the entry barrier to educational and novel radio astronomy projects. The Red Pitaya represents a cost-effective radio signal processing tool that has been considered for both large-scale and small RF projects in radio astronomy (Schleich 2022). They are already heavily used by the Ham radio community. Later Red Pitaya models have a larger bandwidth. We use the STEMLab 125-10 (now discontinued) to show that even the most basic miniature backend can still produce interferometric results.

11.4.1 Description of Red Pitaya Setup

The procedure here utilized Ubuntu 22.04.4 LTS (GNU/Linux 5.15.0-xilinx armv7l) and the Red Pitaya GNU/Linux Ecosystem, Version: 2.00-aff683518, Build: 35 installed on an SD card onto the Red Pitaya. We use the Red Pitaya's spectrometer in place of the ROACH-1, and dump NPZ files onto the control computer. Data could also be written directly onto the SD card but these fill up rapidly. We utilize a 4 TB hot-swappable hard drive on the control computer. The interface between the control computer requires two scripts, one from the control computer that awaits packets from the Ethernet connection to the Red Pitaya, and the other run from the Red Pitaya's OS which takes data as specified.[3] The procedure requires running a Python script from the control computer's command line, and, after logging to the Red Pitaya's OS, running the Red Pitaya script from its command line. The Red Pitaya is shown Figure 11.12, from the Hillsdale College LoFASM system. Note that no Valon or 1 PPS clock are necessary for this experiment, though if one wanted stable time-tagging in the long term, a NetClock or OctoClock model would need to be acquired. These are priced at at least $3000 USD. For a single day of interferometry as in this experience, only the Red Pitaya and control computer are necessary.

[3] https://github.com/tdolch/Radio-Telescope-Instrumentation-for-Teaching

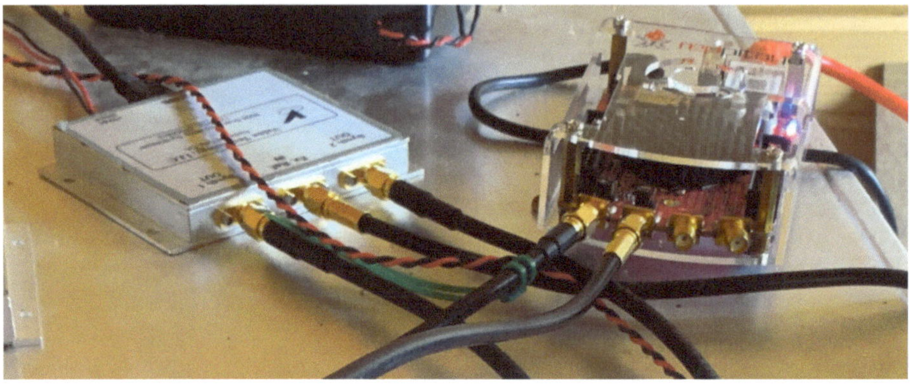

Figure 11.12. Red Pitaya. The two SMA cables attached to the the Red Pitaya's IN1 and IN2 ports go directly to the FEEs on the antennas.

11.5 Using the Red Pitaya Setup for a Low-Cost "LoFASM-Lite"

11.5.1 Results from Demonstration Run

At Hillsdale LoFASM station, we show that the interferometry experiment works in a high-RFI environment near power lines. The outrigger and the main ring are approximately 100 m apart from one another. Each is separated from the shed with the control computer and other hardware by a similar distance. Data were taken for approximately 10 hours. The results are shown in Figures 11.13 and 11.14. Vertical, persistent RFI can be seen, short wave and CB in lower bands, as well as vertical power line noise bars. Additionally horizontal noise transients can be seen. However, the fringes from several astrophysical sources are nonetheless apparent. The difficulty for a student is finding the fringes. The raw data product, after being stacked and downsampled in Python, is a 2D complex array, representing the complex correlations between the two antennas (considering the main ring as a single antenna). The fringes, in this case visible in the real part, must be found by finding the correct color stretch, which is often a very small part of the overall dynamic range, and is challenging to find in the presence of (also) dynamic RFI. Then the delay plot is taken by using the `ifft` command in numpy to perform a fast Fourier transform (FFT) on the data in the horizontal direction. The delay tracks for each source, corresponding each set of fringes, is not immediately obvious. It often lies toward the edge. For a cleaner delay plot, simply take a window from within the fringe plot that minimizes horizontal and vertical RFI. Then transform the subset. The result will have arbitrary delay units but the delay tracks will eventually emerge. The wider baseline one has, one would expect to see fringes in both the real and imaginary parts of the complex correlations.

11.5.2 Results from 2023 Eclipse

During the 2023 and 2024 North American solar eclipses, LWA antenna installations in the DLITE configuration (Chapter 10) were used to image the Sun's corona from the Ohio DLITE station (Young et al. 2025). Earlier, the 2023 North American annular

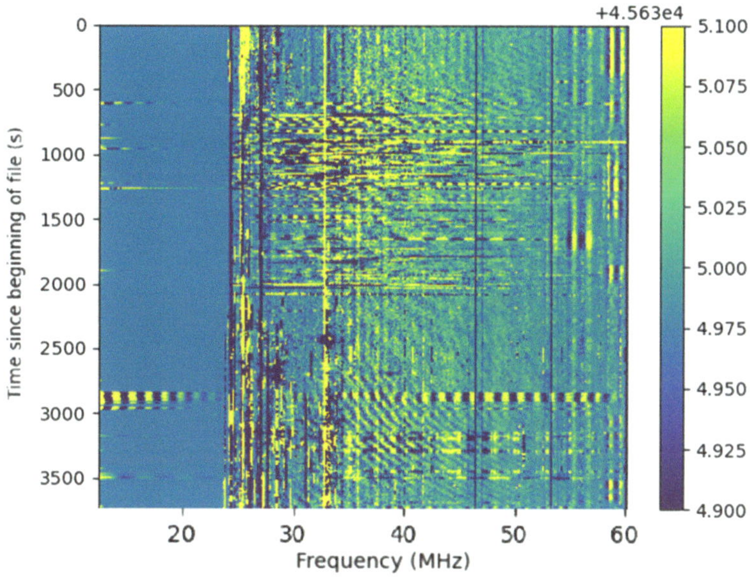

Figure 11.13. Dynamic cross-correlation between main 6-antenna LoFASM ring and single outrigger antenna 100 m away at LoFASM V in Hillsdale, MI, using the Red Pitaya backend receiver. Despite the small base and significant RFI, fringes from the A-team are visible. Intensity units are arbitrary. Note the fine color scale that must be utilized to see the fringe pattern. Each time unit is approximately 10 with the entire spectrograph showing about 10 hr of data.

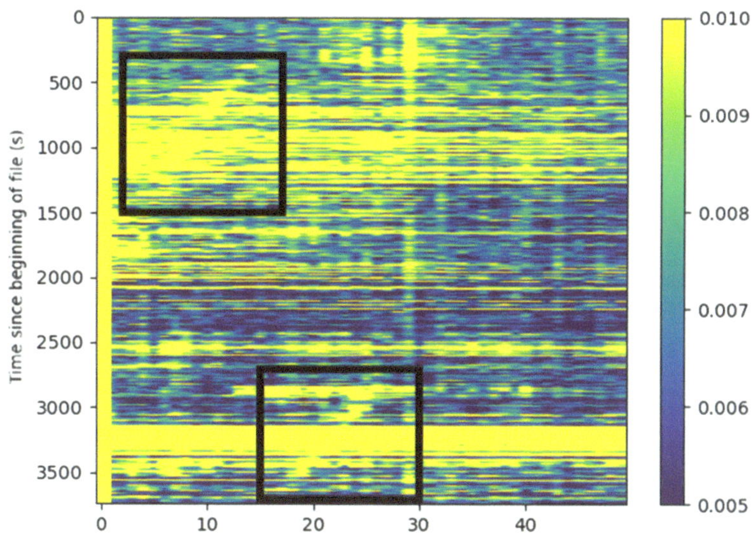

Figure 11.14. Delay from previous figure. Intensity units are arbitrary. The Fourier transform has been taken of the previous plot with fringes, in the horizontal direction. Only the region of the 2D plot in the previous figure has been transformed. The magnitude of the FFT of the real part of the complex correlations between antennas is shown here. The upper box shows Cygnus A, which sets, and the lower box shows Cassiopeia A. Each time unit is approximately 10 with the entire spectrograph showing about 10 hours of data.

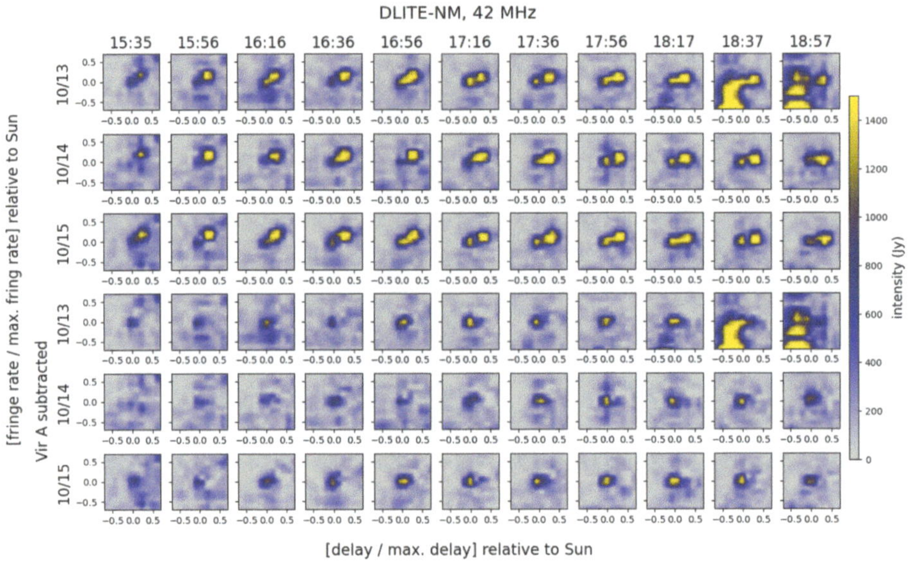

Figure 11.15. The 2023 October 14 annular solar eclipse as viewed from the DLITE station in New Mexico, USA, with only two antennas operating with a baseline of ~350 m. The top three rows show the radio Sun at the center of each square. The off-center source is the "A-Team" radio source Virgo A, corresponding to the galaxy M87. The larger blobs in the last two squares on October 13 are transient RFI. The bottom three rows show Virgo A with a Gaussian fit subtracted to emphasize the eclipsing Sun October 14. Taken together this shows the feasibility of a two-antenna interferometry experiment.

total eclipse's path of totality had crossed the site of the VLA in New Mexico where the LWA1 and the New Mexico DLITE station reside. Solar imaging results from a baseline between a single LWA antenna and another are shown in Figure 11.15. The top rows of images also show Virgo A. This shows that the same point-source experiment with the Red Pitaya at Hillsdale's LoFASM station could be done with only two LWA antennas, if one wished instead to purchase more cable for a longer baseline. While these 2023 eclipse results were not taken with a Red Pitaya, the SDR pipeline is essentially identical in principle to our LoFASM Red Pitaya setup and does not require a faster data rate than what we have already demonstrated in this section. Considerations for which option to choose (only two antennas with a long baseline, or a shorter baseline with a more sensitive ring of six antennas on one end of the baseline) will depend on market costs, terrain, and available student help for assembly. Local RFI near the site of one's radio telescope, though, will not prevent detection of the A-Team sources.

Jupyter notebooks showing the steps used in Python for this analysis are also available.[4] We hope that this experiment, which utilizes FPGA processing, brings students the excitement of "seeing" Cygnus A and Cassiopeia A for the first time. It will parallel the first time one sees the rings of Saturn through an optical telescope. If you have any questions not covered in this text, feel free to contact the editor.

[4] https://github.com/tdolch/Radio-Telescope-Instrumentation-for-Teaching

References

Burrows, D. N., Wolszczan, A., & Moore, A. M. 2021, The WSPC Handbook of Astronomical Instrumentation (Singapore: World Scientific)
Dartez, L. P. 2014, Master's thesis, Univ. Texas
Dartez, L. P. 2021, PhD thesis, Univ. Texas at Arlington
Handzo, E. Y. 2015, Master's thesis, Univ. Texas
Hicks, B. C., Paravastu-Dalal, N., Stewart, K. P., et al. 2012, PASP, 124, 1090
Miller, R. 2012, Undergraduate thesis, Univ. Texas
Price, D. C. 2018, arXiv:1607.03579
Pultinevicius, E., Rockenhäuser, M., Kogel, F., et al. 2023, RScI, 94, 103004
Schleich, S. 2022, MSc thesis, Stellenbosch Univ.
Taylor, C. A., Dowell, J., Taylor, G. B., et al. 2025, JAI, 14, 2550003
Thompson, A. R., Moran, J. M., & Swenson, G. W. 2017, Interferometry and Synthesis in Radio Astronomy (3rd ed.; Berlin: Springer)
Young, O. R., Dolch, T. E., Helmboldt, J. F., et al. 2025, ApJ, 979, 203

Appendix A

Author Biographies

Mojtaba Akhavan-Tafti

Mojtaba Akhavan-Tafti is an associate research scientist at the University of Michigan. His research interests include plasma instabilities in the heliosphere, using observations, simulations, and laboratory experiments. He is the co-investigator and team member of various heliophysics science missions, including the Parker Solar Probe (PSP), the Global Dynamics Constellation (GDC), and the Sun Radio Interferometry Space Experiment (SunRISE) missions. Akhavan-Tafti is also the Principal Investigator of the Space Weather Investigation Frontier (SWIFT) mission concept. SWIFT aims to unravel the three-dimensional structure and temporal evolution of mesoscale solar wind structures that drive terrestrial space weather from inside the Sun–Earth Lagrange point L1 (sub-L1), utilizing flight-ready solar sail technology. He teaches multidisciplinary courses on space instrumentation, as well as space systems design and management, involving hands-on projects in collaboration with government and industry stakeholders.

Joydeep Bagchi

Joydeep Bagchi is an Indian astronomer and currently a Professor at Christ University, Bangalore, India. He received his PhD in Physics from the Indian Institute of Science (IISc). He was a postdoctoral fellow at the Indian Institute of Astrophysics (IIA), Bangalore and Tata Institute of Fundamental Research (TIFR), Mumbai. He was a scientist at the Inter-University Centre for Astronomy and Astrophysics (IUCAA), Pune, for most of his tenure.

He has made significant contributions to the field of astronomy, especially to the field of radio galaxies and large-scale structures. His notable discoveries are the Bullet galaxy and the famous Saraswati Supercluster. While at IUCAA, he set up the Radio Physics Lab in collaboration with the National Centre for Radio Astrophysics, Pune, which carried out various outreach, teaching, training and

doi:10.1088/2514-3433/ae1223ch12

research activities. The lab has been a successful venture and has contributed significantly to spreading astronomy awareness in India and other countries through its activities. The lab under his supervision has trained several students, many of whom are now astronomers.

Kevin Bandura

Kevin Bandura is a Professor at West Virginia University who focuses on building radio telescopes. His research focuses on neutral hydrogen intensity mapping and fast radio bursts and signal processing for radio telescope. He has created the digital signal processing in radio astronomy program to bring radio astronomy engineering into the classroom. He lives in Morgantown West Virginia with his wife and two daughters, who all love exploring the outdoors.

George A. Carson

George A. Carson is a Master's student in astrophysics at Tufts University. He holds a Bachelor of Science in Physics and Mathematics from Dickinson College, where he graduated in 2024. Since 2019, he has contributed to research in solar radio astronomy and remote sensing as an intern at the U.S. Naval Research Laboratory. His work has focused on the analysis of solar radio bursts using data from DLITE, a novel deployable interferometer, as well as on the study of coronal mass ejections through Faraday rotation measurements of extragalactic radio sources. He is a published author in Frontiers in Astronomy and Space Sciences on DLITE's capabilities for solar radio observations.

Sarah Chastain

Sarah Chastain is a postdoctoral researcher studying transients at radio and optical wavelengths. She carries out follow-up observations of gamma-ray bursts using radio telescopes such as the MeerKAT radio telescope, the Australia Telescope Compact Array, and the Very Large Array. Additionally, she conducts searches for any transient phenomena that happens to appear in archival radio observations such as the extensive archive of Long-Wavelength Array all-sky images. Lately she has also been searching for these phenomena at optical wavelengths as well as using observations from the Transiting Exoplanet Survey Satellite.

James M. Cordes

James M. Cordes is the Feldstein Professor of Astronomy at Cornell University. He is also a member of the Cornell Center for Astrophysics and Planetary Science and the Cornell Carl Sagan Institute. His research focuses on many aspects of neutron stars, including the use of pulsars as detectors of long-wavelength gravitational waves and as probes of the interstellar medium. Another primary research area involves fast radio bursts, which also involve neutron stars (magnetars) and perhaps other source classes. Together, pulsars and FRBs provide unique information on interstellar media and the intergalactic medium, particularly small scale turbulence in their ionized gas components. On an observational level, he has explored how propagation effects from interstellar turbulence influence searches for techno-markers from extraterrestrial civilizations. He was involved with the Square Kilometer Array project for ten years and initiated the pulsar survey at the Arecibo Observatory that led to the discovery of the first repeating fast radio burst source.

Louis P. Dartez

Louis Dartez is a postdoctoral scholar at Caltech based at the LIGO Livingston Observatory. His work focuses on gravitational-wave detector calibration and gravitational-wave detector commissioning. He is also actively involved in radio astronomy instrumentation, including development and deployment for projects such as the Low-Frequency All-Sky Monitor (LoFASM). Louis received his PhD in Applied Physics from the University of Texas at Arlington through a cooperative doctoral program with the University of Texas Rio Grande Valley. He was awarded the 2024 LIGO Laboratory Award for Excellence in Detector Characterization and Calibration.

Timothy E. Dolch

Timothy E. Dolch is an Associate Professor of Physics at Hillsdale College. He received his B.S. from the California Institute of Technology and his PhD in Physics & Astronomy from the Johns Hopkins University in 2012. Before joining the faculty of Hillsdale College, he held postdoctoral positions at Oberlin College and Cornell University, both with the North American Nanohertz Observatory for Gravitational Waves (NANOGrav) collaboration. In NANOGrav he has chaired the Education and Public Outreach Working Group. He is also a research scientist with Eureka Scientific and with the SETI Institute. Primarily a transient radio astronomer, his research focuses on pulsars and using them as tools to detect gravitational waves from merging supermassive black holes. He is an author on 98 refereed publications and has taught a broad representation of courses including general relativity, computational physics, electronics, and astronomy. With Hillsdale students, he constructed the Low-Frequency All-Sky Monitor, an on-campus radio telescope, and the Deployable Low-band Ionosphere and Transient Experiment (DLITE) in Ohio.

Jayce Dowell

Jayce Dowell is part of the research faculty at the University of New Mexico and a staff scientist for the Long-Wavelength Array (LWA). Focused on the intersection of science and computing, he works on the design, development, and commissioning of low frequency radio telescopes. Jayce is also the lead developer of the Bifrost high-throughput computing framework and the LWA Software Library, two tools he uses to investigate a range of time domain phenomena, such as pulsars and fast radio bursts.

Joseph Helmboldt

Joseph Helmboldt is a Radio Astronomer at the U.S. Naval Research Laboratory (NRL) in Washington, DC. He received his B.S. in Physics from Central Michigan University in 1999 and his PhD in astronomy from New Mexico State University in 2005. He was a postdoctoral fellow under Professor Greg Taylor at the University of New Mexico for one year following graduate school. After this, he was a National Research Council fellow at NRL for three years before being hired there as a civil servant. While Joe has cycled through several areas of study, including optical studies of extra-galactic star formation and very long baseline interferometry of active galactic nuclei, he has developed expertise in the use of radio astronomical techniques to study the dynamics of Earth's ionosphere. Within this field, he has pioneer several methodologies and was the lead developer of NRL's Deployable Low-band Ionosphere and Transient Experiment (DLITE).

Chuck Higgins

Chuck Higgins grew up in Huntsville, Alabama, and the rumblings of the Apollo engine test firings in the 1970s at the nearby NASA Marshall Space Flight Center captured his interest in astronomy and space. He completed a B.S. in Physics from the University of Alabama-Huntsville and received M.S. and PhD degrees from the University of Florida where he studied the radio emissions from the planet Jupiter. He was a National Research Council postdoctoral fellow at NASA's Goddard Space Flight Center and is a founding member of Radio JOVE, a NASA-partner citizen science and education project. He is a Professor of Physics and Astronomy at Middle Tennessee State University.

Bhal Chandra Joshi

Bhal Chandra Joshi obtained his Bachelor of Engineering degree from Indian Institute of Technology, Roorkee (India) and Master of Technology from Indian Institute of Technology, Powai (India). He subsequently was awarded Doctor of Philosophy from University of Pune in astronomy and physics. He was involved in setting up the Giant Meterwave Radio Telescope (GMRT) from its early days. He contributed to the design, analysis and commissioning of the servo system of the GMRT antennas and in developing the hardware and software for the pulsar receivers at the GMRT. He co-founded and led for more than a decade the Radio Physics Laboratory (RPL) at the National Centre for Radio Astrophysics (NCRA), Pune (India), where he designed and developed several low-cost radio astronomy instruments and experiments as teaching aids for radio astronomy to college students. Over the years, the RPL has held a radio astronomy winter school every year training more than a thousand students. Apart from developing radio astronomy instrumentation, he is an active pulsar astronomer. His research interests include search for new pulsars as well as studies of their single pulses. A survey led by him discovered the only known double pulsar system. In the last decade, he has been associated in setting up precision pulsar timing with the the GMRT. He established the Indo-Japanese Pulsar Timing Array (InPTA) experiment with the GMRT and led it for over a decade. Currently, he holds the Raja Ramanna Chair at NCRA, where he keeps pursuing his radio astronomy research.

Jason E. Kooi

Jason E. Kooi works at the U.S. Naval Research Laboratory as a research physicist in the Remote Sensing Division. Seemingly unable to pursue "Ghostbusting" as a major in college, he, instead, completed a B.S. in Physics with a minor in History and a B.S. in Math at New Mexico Tech. He then completed a M.S. in Math Analysis there before moving to the University of Iowa to complete a PhD in Physics, studying Faraday rotation induced by the solar wind and coronal mass ejections. His current research interests include the solar wind, space weather, and machine learning.

Glen I. Langston

Glen I. Langston works for the U.S. National Science Foundation as a program officer in the Astronomy Division. He leads proposal reviews for topics concerning the structure of the Milky Way, Galactic Stellar Clusters and the Interstellar Medium. Glen reviews, and recommends funding for, proposals concerning Astronomical Education and Special Programs. Glen is a former research scientist at the National Radio Astronomy Observatory and was project scientist for a number space craft radio astronomy projects. He lives with Katherine, his wife, in Green Bank, WV and continues to operate an Array of Horn Radio Telescopes in his back yard. Glen is also an Elder for the Liberty Presbyterian Church.

Ashish A. Mshake

Ashish Mhaske is a Marie Curie Fellow at the Kavli Institute of Cosmology, Cavendish Laboratory, University of Cambridge. He has worked in the field of astronomy and instrumentation at the Inter-University Centre for Astronomy and Astrophysics (IUCAA), India. He previously worked in the Laser Interferometer Gravitational-Wave Observatory (LIGO) - India project on experimental aspects of gravitational wave detectors. His current work at Cambridge spans space based radio interferometers and the Square Kilometer Array instrumentation and science.

Dhruba J. Saikia

Dhruba J. Saikia was a Professor at the National Centre for Radio Astrophysics of the Tata Institute of Fundamental Research (TIFR) where he worked for over 40 years, and was the Founding Vice-Chancellor of Cotton University at Guwahati. After his superannuation from TIFR, he headed the Astronomy Centre for Educators of the Inter-University Centre for Astronomy and Astrophysics

(IUCAA). His research interests are on different aspects of extragalactic astronomy and astrophysics, especially active galaxies, and more recently on education, with an emphasis on higher education. Besides TIFR and IUCAA, he has been engaged in research and/or teaching at the Jodrell Bank Observatory of the University of Manchester; National Radio Astronomy Observatory USA; Queen's University at Kingston, Canada; Australia Telescope National Facility (CASS) of CSIRO Australia; the University of Western Australia and Savitribai Phule Pune University and Cotton University, India. He has published over 180 research papers in leading journals, over a 110 in conference proceedings, and co-edited three volumes: Low-Frequency Radio Universe published by the Astronomical Society of the Pacific in 2009; Fluid Flows to Black Holes, A tribute to S Chandrasekhar on His Birth Centenary published by World Scientific in 2013; and Selected Progresses in Modern Physics, Proceedings of TiMP 2021, published by Springer in 2021. He is an Editor of Astrophysics and Space Science, published by Springer. He has contributed to the Indian Higher Education Report 2023 from the Centre for Policy Research in Higher Education, National Institute of Educational Planning and Administration (NIEPA), published by Routledge. He was awarded the Professor M K Vainu Bappu Gold Medal of the Astronomical Society of India and is a Fellow of the National Academy of Sciences, India. Among his varied interests he is also a keen birder.

Shirsh L. Soni

Dr. Shirsh Lata Soni is an early-career researcher dedicated to the field of solar physics. Her research interests entail conducting fundamental scientific research in solar physics and space weather using observations and modeling, as well as contributing to the planning and designing of heliophysics missions. She is dedicated to mentorship and outreach, guiding high school and undergraduate students in space science education through NASA-sponsored programs.

Gregory B. Taylor

Gregory Taylor is a Distinguished Professor in the department of Physics and Astronomy at the University of New Mexico, where he is also the Founding Director of the Long-Wavelength Array and the Director of the Center for Astrophysical Research and Technologies. Before that he spent over 10 years on the scientific staff at the National Radio Astronomy Observatory in Socorro, New Mexico. Professor Taylor's research interests include meteors, space weather, pulsars, radio afterglows from gamma-ray bursts, and active galaxies and their environments. He is a member of the IAU, the AAS, and is a past member of the Committee on Radio Frequencies, a standing committee of the National Academy of Sciences. He has been the principal advisor for 12 PhD students. Professor Taylor has co-authored or edited three books and over 300 refereed publications.

David Typinski

Dave Typinski is a professional businessman and amateur scientist who has been tinkering with things electrical and mechanical since he was old enough to hold a soldering iron and a Crescent wrench. His primary scientific interests are radio astronomy, mathematics, and the history of technology. Dave is a licensed amateur radio operator, call sign AJ4CO. He operates AJ4CO Observatory (aj4co.org) and is an active member of the Radio Jove project, serving as part of its management group. Dave lives in High Springs, Florida with his wife and cats, all of whom graciously pretend he is normal.

www.ingramcontent.com/pod-product-compliance
Ingram Content Group UK Ltd.
Pitfield, Milton Keynes, MK11 3LW, UK
UKHW061825120926
8556IPUK00035BA/929
* 9 7 8 0 7 5 0 3 5 6 0 1 5 *